Biodiversity in the North West: The Slime Moulds of Cheshire

Biodiversity in the North West: The Slime Moulds of Cheshire

Bruce Ing

University of Chester Press

First published 2011
by University of Chester Press
University of Chester
Parkgate Road
Chester CH1 4BJ

Printed and bound in the UK by the
LIS Print Unit
University of Chester
Cover designed by the
LIS Graphics Team
University of Chester

A catalogue record for this book is available from the British
Library

ISBN 978-1-905929-91-7

CONTENTS

ACKNOWLEDGEMENTS

I wish to thank all the individual collectors who have made their collections and notes available to me, especially Terry Palmer, members of the North West Fungus Group and past students of Chester College (now the University of Chester). I would like to record my thanks for their support and encouragement over many years to past and present academic, technical and library staff at the University of Chester, and its earlier incarnations. Finally I must pay tribute to my long-suffering wife, Eleanor, who has endured more than four decades of marriage to a myxomycologist, with all that that entails!

LIST OF ILLUSTRATIONS

List of Illustrations

LIST OF ABBREVIATIONS

BM – Herbarium of the Natural History Museum, London
CMG – Herbarium at the University of Cambridge
E – Herbarium at the Royal Botanic Garden, Edinburgh
G – Herbarium of the Conservatoire and Botanical Garden of the
    City of Geneva
K – Herbarium at the Royal Botanic Gardens, Kew
LIV – Herbarium at the Liverpool Museum
LIVU – Herbarium at the University of Liverpool
MANCH – Herbarium at the Manchester Museum
MANX – Collection at the Manx Museum, Douglas

PART 1 – INTRODUCTION

1.1   The biology of slime moulds

The organisms collectively known as slime moulds share some fungal and some animal, or rather, protozoan characteristics. They have amoeboid and/or flagellate single-celled feeding stages and usually some form of aggregation into either a multicellular structure – the *pseudoplasmodium*, or a multinucleate single cell – the *plasmodium*, which migrates from the feeding area of the substrate to a region where it can produce spores.

There are five major groups of slime moulds – protostelids, dictyostelids, ceratiomyxids, myxomycetes and acrasians. The first four are branches of the same evolutionary line while the acrasians have developed from a quite different amoeboid ancestor. All are now recognised as members of the kingdom Protozoa but have been classified with the fungi in the past. They have been called 'fungus-animals' or Mycetozoa, a term which is coming back into fashion.

The protostelids are the smallest and simplest, some have flagellate cells and others do not. They mostly have a small multinucleate plasmodium which, in some cases, is the result of a sexual process. The fruiting body is usually a thin, hollow stalk surmounted by a cluster of a few spores. Protostelids live on the outside of dead plant stems, grass flower-heads and the bark of living trees. They can only be seen when these substrates are cultured in moist chambers. The group has been little studied in Britain and has not so far been found in Cheshire.

Dictyostelids, or cellular slime moulds, are probably derived from non-flagellated protostelids but differ in several important ways. Neither sexual reproduction nor a

1

plasmodium occur, instead individual amoebae move towards the source of a chemical attractant, *acrasin*, and form the aggregation phase or *pseudoplasmodium*. This transforms into a slug-like *grex* and moves towards a drier and higher part of the substrate. Spore production involves a division of labour in which a cellular stalk is formed from some of the cells while those which secrete the most acrasin crawl up to the top of the stalk and form resting spores. Dictyostelids are primarily soil organisms but are also common on herbivore dung. As with all the other groups of slime moulds they feed by ingesting bacteria, other protozoans, algae or yeast fungi. They are perhaps more common in the tropics than in temperate regions and have received only moderate attention in Britain, and very little in Cheshire.

Ceratiomyxids are closely related to the protostelids but differ in having massive plasmodia, often a few centimetres in diameter, a well-developed sexual phase, with fusion of compatible cells to form the plasmodium, and a fruiting structure which is conspicuous to the naked eye, consisting of finger-like processes carrying very small sporangia, which in the past were interpreted as spores. They have traditionally been included with the myxomycetes. There are five species of the single genus, *Ceratiomyxa*, of which three occur in Britain, but only the most common, *C. fruticulosa*, is known in Cheshire. This is found on rotten wood, especially of conifers and is one of the most common and widespread species in the world.

The myxomycetes, or plasmodial slime moulds, are much more complex. The plasmodial stage is often large and brightly coloured and is capable of visible movement across the substrate. The fruit bodies often resemble miniature mushrooms, usually only a millimetre or so high, and may also be brightly coloured. As a group they occur in all climates and in all kinds of vegetation, from deserts to alpine snowfields but are most common in forests, occurring on fallen wood of all kinds, leaf litter, plant remains, mosses, stems of living plants and the bark of living trees. The last habitat contains a wealth of minute species barely visible to the naked eye and usually only found by culturing pieces of bark from living trees in moist chambers – usually Petri dishes. Many species are cosmopolitan while a few are confined to the tropics and a larger group are only known from vegetation close to melting snow on high mountains in the spring. There are probably as many as one thousand species of myxomycetes but only about 700 have been described to date. They are well researched in Britain, with over 400 species recorded, and also well studied in Cheshire – the bulk of the slime moulds included here are myxomycetes.

The acrasians are not closely related to other slime moulds, having quite different amoeboid stages. A flagellate stage is reported from one species. Sexual reproduction has not been observed. The plasmodial stage is inconspicuous. Only one species is common, *Pocheina rosea*, which grows on acid bark. This has a stalk made up of cubic cells but could be mistaken for a species of the myxomycete genus *Echinostelium* which also lives on bark but does not have a cellular stalk, just a hollow tube. Other acrasians are found on dung, plant litter or old wine casks, but are generally regarded as rare. It is more likely that they are just insufficiently studied, especially in Britain.

Myxomycetes are abundant on bark, even in cities, and in soil of all kinds, including deserts and coastal dunes, where they may be important agents in the phosphate cycle. They are especially common on dead stems of nettles and rosebay willow herb which are typically found in phosphate-rich soils. Forest litter is also rich in species and the fruit bodies may be found at the litter surface. In desert regions of the Americas and Africa rotting cacti and other succulents carry a characteristic suite of myxomycetes. Several of these now occur on naturalised cacti, succulent spurges and *Agave* species in the Mediterranean region and the Canary Islands. Dead wood is an important habitat and some of the larger species may feed inside a fallen trunk for several years before emerging, often through a beetle bore-hole, to produce spores at the surface. Spores are mostly wind dispersed but some species appear to have formed a partnership with certain beetles, whose complete life-cycle occurs within the large clusters of fruiting structures on dead wood. The female insects carry spores in pouches in their jaws and have been observed 'sowing' spores on rotten wood. For more information on the biology and identification of myxomycetes see Ing (1994, 1999 and 2008).

1.2 The county of Cheshire and its natural characteristics

Which area is included?

The area under consideration is the Watsonian vice-county 58 Cheshire, which consists of the County Palatine, including the whole of the Wirral Peninsula. Modern 'Cheshire' has lost Wirral, those parts of the county south of the Mersey which are part of Greater Manchester and Stockport, and parts of the Pennine area now in Derbyshire, notably the 'panhandle' of Longdendale. All these areas of 'old' Cheshire are included here. Present-day 'Cheshire' now includes the boroughs of Widnes and Warrington, both formerly in Lancashire and part of vice-county 59, South Lancashire. As biological recording now carried out in Cheshire incorporates these ex-Lancashire areas they, too, are included in this account. Even more

4

recently Cheshire has been split into east and west authorities. These, often ephemeral, changes entirely justify the use of the stable unit of the vice-county, which will not change. Map 1 on page 59 shows the area, its 10 km squares and the surrounding vice-counties.

The physical and ecological environment

The geology of the area is relatively simple. The greater part consists of Triassic sandstones with prominent ridges of harder sandstones – Bunter pebbles and Keuper waterstones – along Wirral and in the main ridge from Runcorn southwards and at Alderley Edge. The plain between these ridges is covered with post-glacial deposits, mostly boulder clay. In the east, on the Pennine edge, the red sandstones are replaced by Carboniferous grits and shales, especially Millstone Grit. The low-lying areas have numerous depressions and kettle-holes which form deep, steep-sided meres. Many of these have become filled in with peat-based vegetation – these are the 'mosses' – and several have been lost to drainage. The best examples are Wybunbury Moss, Abbots Moss, Flaxmere and Hatchmere, the last two in Delamere Forest. The surface of *Sphagnum* mosses is an important substrate for a wide range of specialised organisms, including slime moulds. The plain is largely covered in grazing land, often damp pasture, with moorland on the eastern grits and lowland heath on the sandstone ridges, now much overgrown with birch/pine scrub. Many of the river valleys have cut deep valleys, or cloughs, and their steep banks are clothed in small amounts of ancient woodland, such as Cotterill Clough and Styal Woods. The sandstone ridges also have some semi-natural woodland, such as the remnants of the old Wirral forest at Eastham, and sizeable conifer plantings, notably in Delamere Forest. Many of the large estates on the plain have extensive amenity woodland and parkland, such as Eaton Park, Chester; Dunham Massey; Marbury Park; and Tatton Park. The great sweep of Lyme Park has remnants of Pennine sessile oak forest, acid grassland and moorland on the Grit. Along the major rivers extensive marshland has developed and along the Dee estuary there is a considerable area of saltmarsh. At the top of Wirral there are several areas of sand dunes, which although much altered for recreation, are still important for many plants, fungi and slime moulds.

1.3 The study of slime moulds in Cheshire

The earliest collector of slime moulds, among other fungi, lichens and mosses, in Cheshire was W. Wilson, who was active between 1827 and 1854. In 1859 the Rev. H.H. Higgins published a list of fungi, including some myxomycetes, from the Liverpool region, including parts of Cheshire, and especially Wirral. In 1872 a Mr T. Brittain collected in Chester what was to become the type specimen of *Didymium tussilaginis*, now in the herbarium of the Natural History Museum, London (**BM**). In 1895 the then world authority, Arthur Lister, collected a few specimens in the neighbourhood of Wilmslow. These are listed in the Lister notebooks, which, together with the specimens, are in **BM**.

W.H. Pepworth collected a few species between 1896 and 1900 and H. Murray did likewise between 1899 and 1904. Their material is in the herbarium at Manchester Museum (**MANCH**). Around 1906 J.R. Hardy made a few collections. During the period 1909 to 1915 the distinguished amateur mycologist Dr J.W. Ellis collected intensively on Wirral and published a valuable paper (Ellis, 1915). His material is in the herbarium at the Royal Botanic Gardens, Kew (**K**) and **BM**.

Part 1 – Introduction

In the 1920s the well-known Lancashire botanist W.G. Travis collected in Cheshire.

The renowned amateur mycologist W. Douglas Graddon, who specialised in ascomycetes, lived at Congleton. He had been trained in the study of myxomycetes in his earlier home area of Epping Forest, Essex, by Gulielma Lister (daughter of Arthur), the world authority of her day. He collected extensively in the Pennine areas of the Congleton and Macclesfield districts, and over the border into Staffordshire, between 1941 and 1951, and his card index and collections of myxomycetes are in the present author's herbarium. S.S. Bates, who lived at Stalybridge before retiring to Bradford-on-Avon, collected widely in the north east of the county between 1951 and 1964. His material is now at **K**.

J.T. Palmer, a mycologist of international distinction, is a specialist in gastroid basidiomycetes and ascomycetes but made several valuable records of myxomycetes between 1953 and 1963. His collections were originally in the herbaria of the University of Liverpool (**LIVU**) and Liverpool Museum (**LIV**) but are now at **K**. His notebooks have been studied by the present author. J.E. Milne, who lived at Bramhall before retiring to the Isle of Man, collected extensively in the area south of Manchester, between 1960 and 1964. His collections are in the Manx Museum, Douglas (**MANX**) and at **K**.

The Liverpool Botanical Society, the Liverpool Field Naturalists, the Merseyside Naturalists, the North West Naturalists' Association and, more recently, the North West Fungus Group have held many forays over the years and several members have contributed valuable myxomycete records, notably Beth Hartham, Ken Jordan, Adrian Newton, Ron Poole and Gordon Taylor.

The present author came to teach at Chester College of Education (now the University of Chester) in 1971 and has collected throughout the county up to 2011. Several students have been encouraged to study slime moulds and of these Andy Ryan, of Newton, Wirral and Diana Wrigley de Basanta, now living in Madrid, have made important contributions to our knowledge of myxomycetes far beyond Cheshire, while others have confined their research activities to the environs of Chester. All their identifications have been confirmed by the present writer.

Cheshire has been fortunate to have had the university cities of Liverpool and Manchester, now joined by Chester, close by, which has allowed enthusiastic mycologists to foray in the county, especially in 'honey-pot' spots such as Delamere Forest. The work of the local societies mentioned above has been supplemented by a series of residential forays organised by the British Mycological Society in 1910 (Chester), 1965 (Manchester), 1972 (Liverpool) and 1985 (Chester). These have added many records to the Cheshire list. In 1988 the author ran a workshop on myxomycetes at Chester, for the Society, which also added good records. The first International Congress on the Systematics and Ecology of Myxomycetes was held at Chester in 1993. In recent years mycological activities have centred on the North West Fungus Group which has a comprehensive programme of field meetings and these have produced more records. During the period 2007 to 2011 the author has visited every 10 km grid square in the county to collect field samples and bark, for moist chamber culture. Map 2 on page 60 shows all the 5 km grid squares in which at least one species has been reliably recorded.

In the account of Cheshire slime moulds in Part 2 notes are given on the occurrence of species in the counties, or, more accurately, vice-counties surrounding Cheshire. All are well studied but the following table provides an indication of their comparative richness, as known at present.

Table 1: Occurrence of slime moulds in the vice-counties surrounding Cheshire.

County	Vice-county	Number of species
Cheshire	58	182
Denbighshire	50	176
Shropshire	40	174
Derbyshire	57	168
Flintshire	51	158
South Lancashire	59	151
South West Yorkshire	63	145
Staffordshire	39	119

Coverage of the area has been fairly good, with the number of species recorded in a 10 km grid square more or less proportional to the amount of that square which lies in Cheshire, as understood here. The following table allows some comparisons, the poorer squares being largely devoid of woodland, or having little of Cheshire.

The remaining fragments of squares in Cheshire have little or no suitable habitat and have no recorded myxomycetes.

The distribution maps of species show clearly the sparse occurrence of those species which sporulate on fallen wood or forest leaf litter in the southern area of the Cheshire Plain, where native woodland is rare and where the estate woodlands are not always accessible to collectors. In contrast, the species growing on the bark of living trees are found throughout the county, on roadside and parkland trees as well as in natural woodlands.

Table 2: Comparison of 10 km squares for incidence of slime moulds.

100 km square	10 km square	No. of species	100 km square	10 km square	No. of species
SJ	27	24	SJ	67	58
	28	58		68	71
	29	28		69	32
	35	23		74	22
	36	39		75	44
	37	61		76	50
	38	64		77	53
	39	18		78	76
	44	35		79	22
	45	64		85	24
	46	85		86	53
	47	58		87	56
	48	20		88	76
	54	39		89	19
	55	62		96	23
	56	78		97	51
	57	99		98	56
	58	55		99	84
	59	21	SD	90	32
	64	39	SK	09	48
	65	61	SE	00	21
	66	56			

1.4 List of individual collectors

W.B. Allen, S.S. Bates, D. Bell, T. Brittain, J.W. Ellis, W.D. Graddon, J.R. Hardy, B. Hartham, A. Hodgkinson, N. Hughes, B. Ing, E.B. Ing, K. Jordan, A. Lister, J. Ley, J.E. Milne, H. Murray, A. Newton, J.T. Palmer, W.H. Pepworth, R. Poole,

S. Runagall, A. Ryan, P.R. Smith, G. Taylor, W.G. Travis, J.T. Wadsworth, W. Wilson, G. Wimpey and D. Wrigley-Basanta.

Where a collector's name does not appear in the species accounts his/her collections were of more common species. The initials *BMSF* and *NWFG* refer to collections made during British Mycological Society forays and those of the North West Fungus group, respectively.

1.5   Sources of records

Records have been collated from a variety of sources. The list of references in the bibliography gives details of published accounts. The herbaria at **BM**, University of Cambridge (**CMG**), Royal Botanic Garden, Edinburgh (**E**), the Conservatory and Botanical Garden of the City of Geneva (**G**), **K**, **MANCH** and **MANX** have been studied. The author's extensive herbarium has voucher specimens for most of the species recorded in Cheshire.

The Lister notebooks, owned by the British Mycological Society, but held in the Botanical Library at the Natural History Museum in London, have a wealth of information, including some early Cheshire records. The foray database of the British Mycological Society has details from all its forays, especially useful for those where the results were not published separately. The foray records database of the North West Fungus Group has also provided much information.

Finally, the account would not be complete without the records provided by the individual collectors named in the previous section.

1.6 Layout of entries in the systematic account

Currently accepted name and author citation
Synonyms in current literature
Habitat of species in Cheshire; general frequency in Britain.
Where there are fewer than six records: the localities (in chronological order), years and names of collectors. Where there are six or more records: year of first record, year of last record. All 10 km grid squares (in Cheshire) in which the species has been reliably recorded. Number of individual sites (separated by at least one kilometre) in which the species is recorded. (Maps are provided for those species with six or more records.) Vice-counties surrounding Cheshire in which the species is recorded. Any specific notes on the ecology or distribution of the species. Number of map where appropriate.

The distribution maps are based on 5 km squares of the National Grid – open circles indicate pre-1960 records and closed black circles indicate post-1960 records.

The nomenclature follows Ing (1999).

1.7 Bibliography

Ellis, J.W. (1915) The Mycetozoa of Wirral. *Lancashire and Cheshire Naturalist* **8**, 2–5.

Higgins, H.H. (1859) The Fungi of Liverpool II. Gasteromycetes. *Proceedings of the Literary and Philosophical Society of Liverpool* **1859**, Appendix, 123–138.

Holden, M. (1966) Autumn Foray, Manchester, 3–10 September 1965. *British Mycological Society News Bulletin*, **26**, 1–4.

Ing, B. (1994) The Phytosociology of Myxomycetes. *New Phytologist* **126**, 175–201.

Ing, B. (1999) *The Myxomycetes of Britain and Ireland.* Slough, Richmond Publishing.

Ing, B. (2003) *Licea margaritacea,* a new Myxomycete from Sycamore Bark. *Mycologist* **17**, 28–29.

Ing, B. (2008) *The Exciting World of the Slime Moulds.* Inaugural and Professorial Lecture. Chester, Chester Academic Press.

Lister, G. (1911) *Monograph of the Mycetozoa* 2nd ed. London, British Museum.

Massee, G. (1892) *Monograph of the Myxogastres.* London, Methuen.

Rea, C. (1911) The Chester Spring Foray, 13–17 May 1910, *Transactions of the British Mycological Society* **3**, 233–238.

Smith, A.L. & Rea, C. (1905) Fungi New to Britain. *Transactions of the British Mycological Society* **2**, 92–99.

Thomas, A. (1973) Autumn Foray, Liverpool, September 1972. *Bulletin of the British Mycological Society* **7**, 52–58.

PART 2 - SYSTEMATIC ACCOUNT

THE SLIME MOULDS OF CHESHIRE

Phylum MYXOMYCOTA
Class ACRASIOMYCETES
Order ACRASIALES
Family Guttulinaceae

Pocheina rosea (Cienk.) A.R. Loeblich & Tappan
Guttulina rosea Cienk.
On acid bark of living trees, in moist chamber culture; common, especially in urban areas affected by atmospheric pollution.
First record 1977, last record 2010.   SJ 36-38, 45-47, 56-58, 64-69, 74-77, 79, 87, 89, 97. 25 sites.   An albino form has been found at a few sites. (Map 3) Recorded in all neighbouring counties.

Guttulinopsis vulgaris E.W. Olive
On wild rabbit dung, in moist chamber culture; rarely recorded in Britain.
Eaton Park, Chester, 1977, *J. Ley.* SJ 46. Not known in surrounding counties.

Class DICTYOSTELIOMYCETES
Order DICTYOSTELIALES
Family Dictyosteliaceae

Dictyostelium brefeldianum Hagiwara
Dictyostelium mucoroides auct.
On dung of domestic and wild rabbits, in moist chamber culture; common.
Eaton Park, Chester, 1977 and Peckforton, 1977, *J. Ley*; Chester, 1987, *D. Bell*.   SJ 46, 55.   Recorded from Denbighshire, Flintshire and South Lancashire.

Dictyostelium giganteum Singh
On domestic rabbit dung, in moist chamber culture; frequent.
Chester, 1987, *D. Bell*.   SJ 46.   Recorded from Denbighshire and Flintshire.

Dictyostelium lacteum Raper
On domestic rabbit dung, in moist chamber culture; uncommon.
Chester, 1987, *D. Bell*.   SJ 46.   Not known in surrounding counties.

Class CERATIOMYXOMYCETES
Order CERATIOMYXALES
Family Ceratiomyxaceae

Ceratiomyxa fruticulosa (Müll.) T. Macbr.
On rotten wood, especially of conifers; common.
First record 1941, last record 2010.   SE 00; SJ 36–38, 45–47, 54–58, 64–68, 75–78, 86–88, 97, 99; SK 09.
31 sites. (Map 4) Recorded in all neighbouring counties.

Class MYXOMYCETES
Order ECHINOSTELIALES
Family Echinosteliaceae

Echinostelium brooksii K.D. Whitney
On the acid bark of living trees, in moist chamber culture; common.
First record 1993, last record 2010. SE 00; SJ 28, 36–39, 44–48, 54, 55–58, 64–69, 75–78, 87–89, 97–99.
38 sites. (Map 5) Recorded in all neighbouring counties.

Echinostelium colliculosum K.D. Whitney & H.W. Keller
On the bark of living trees, in moist chamber culture; common.
First record 1979, last record 2010. SE 00; SJ 27, 28, 36–39, 44–47, 54–57, 59, 64–69, 75–78, 85, 87, 89, 97, 99. 38 sites. (Map 6) Recorded in all neighbouring counties.

Echinostelium corynophorum K.D. Whitney
On bark of living trees, in moist chamber culture; frequent.
First record 2006, last record 2010. SJ 29, 45–47, 55, 58, 59, 64–69, 75, 77, 88, 07. 20 sites. (Map 7) Recorded from all surrounding counties except Staffordshire.

Echinostelium fragile Nann.-Bremek.
On bark of living trees, in moist chamber culture; frequent.
First record 1993, last record 2010. SD 90; SJ 29, 35, 36, 44–47, 54–58, 64–69, 75–79, 85, 87, 97–99.
35 sites. (Map 8) Recorded in all neighbouring counties except South West Yorkshire.

Echinostelium minutum de Bary
On the bark of living trees, in moist chamber, rarely on other plant debris; common.
First record 1924, last record 2011. SD 90; SE 00; SJ 27, 28, 36–39, 44–48, 54–59, 64–69, 74–78, 87–89, 96–99; SK 09. 49 sites. (Map 9)   Recorded in all neighbouring counties.

Echinostelium vanderpoelii Nann.-Bremek., D. W. Mitchell, T.N. Lakh. & R.K. Chopra
On the bark of living *Metasequoia*; uncommon.
Walton Hall Gardens, 2010, *B. Ing.* SJ 68. Recorded from Derbyshire. This species is often equated with the much less common *E. apitectum* K.D. Whitney, but it seems to be constantly different in Britain.

Order LICEALES
Family Liceaceae

Licea belmontiana Nann.-Bremek.
On bark of living trees, in moist chamber culture; uncommon.
First record 2007, last record 2011. SJ 45–47, 65, 66, 74, 76, 97, 98.   Nine sites.   (Map 10) Recorded from Derbyshire and South Lancashire.

Licea biforis Morgan
On bark of living trees; frequent, becoming more common as the climate warms.
First record 2007, last record 2010. SD 90; SJ 29, 39, 46, 58, 59, 64, 65, 67, 68, 77, 99; SK 09. 17 sites. (Map 11)   Recorded from Shropshire, Denbighshire, Flintshire, South Lancashire and South West Yorkshire.

Licea bryophila Nann.-Bremek.
On *Metzgeria* and *Radula* on the bark of living trees; frequent in the west.
First record 2008, last record 2010. SD 90; SJ 38, 56, 67, 77, 55. Six sites. (Map 12) Recorded from Denbighshire.

Licea castanea G. Lister
On bark of living trees; uncommon.
First record 1985, last record 2010. SJ 29, 56, 58, 78, 89. Five sites. Recorded from Denbighshire and Derbyshire.

Licea clarkii Ing
On dead bramble stems; frequent.
First record 1989, last record 2010. SD 90; SJ 35, 45–47, 64–68, 76, 77, 87, 99. 16 sites. (Map 13) Recorded from all surrounding counties except South West Yorkshire.

Licea denudescens H.W. Keller & T.E. Brooks
On bark of living trees, in moist chamber; frequent.
First record 2006, last record 2010. SJ 28, 39, 44, 46, 58, 66, 68, 69, 75, 77, 89, 97. 15 sites. (Map 14) Recorded from Denbighshire, Flintshire and South Lancashire.

Licea eleanorae Ing
On bark of living trees; uncommon.
First record 2007, last record 2011. SD 90; SJ 27, 29, 36, 39, 44, 47, 55, 58, 64–66. 13 sites. (Map 15) Recorded from Denbighshire, Shropshire and South Lancashire. This species, which was originally described, in 1999, from Switzerland, is named after the author's wife.

Licea erddigensis Ing
On bark of living trees, in moist chamber; uncommon.
First record 2006, last record 2010. SJ 27, 29, 47, 48, 58, 65, 66, 77, 99. 11 sites. Described from Erddig Park, near Wrexham, in 1999. (Map 16) Recorded from Shropshire, Denbighshire and South Lancashire.

Licea inconspicua T.E. Brooks & H.W. Keller
On bark of living trees, in moist chamber; uncommon.
First record 1986, last record 2010. SJ 36, 67, 79, 85, 87. Five sites. Recorded from Shropshire, Denbighshire, Flintshire and South Lancashire.

Licea kleistobolus G.W. Martin
On the acid bark of living trees, in moist chamber; common.
First record 1997, last record 2011. SD 90; SE 00; SJ 27–29, 35–39, 44–48, 54–58, 64–69, 74–79, 85–89, 96–99; SK 09. 74 sites. (Map 17) Recorded in all neighbouring counties.

Licea longa Flatau
On mossy bark of living oak tree; uncommon.
Broxton, 2010, *B. Ing*. SJ 45. Recorded from Derbyshire and Flintshire.

Licea margaritacea Ing
On bark of living sycamore and willow trees; rare.
Mossley, 2008 and Whitby Park, Ellesmere Port, 2010, *B. Ing*. SD 90; SJ 37. Described recently (Ing, 2003) from Dumfries this species is otherwise known from one further site in Scotland, one in Wales and one other in England; it has been found once in Switzerland.

Licea marginata Nann.-Bremek.
On bark of living trees; frequent.
First record 2007, last record 2010. SD 90; SJ 39, 46, 55, 58, 64, 66–69, 74–78, 86, 99. 20 sites. (Map 18) Recorded from neighbouring counties except Staffordshire and South West Yorkshire.

Licea microscopica D.W. Mitchell
On bark of living elder trees, in moist chamber; common.
First record 2003, last record 2011. SD 90; SJ 27–29, 35–38, 44–48, 54–59, 64–69, 74–79, 85, 87, 89, 96–99; SK 09. 55 sites. (Map 19) Recorded from Shropshire, Denbighshire, Flintshire and South Lancashire.

Licea minima Fr.
On fallen conifer logs and the bark of living trees, in moist chamber; common.
First record 1965, last record 2010. SJ 28, 36, 45, 55–58, 65, 67, 68, 78. 14 sites. (Map 20) Recorded in all neighbouring counties.

Licea operculata (Wingate) G.W. Martin
On bark of living trees; frequent, possibly increasing in Britain.
First record 2007, last record 2011. SD 90; SJ 36, 44, 46–48, 54, 55, 58, 64–67, 58, 74–76, 78, 89, 96, 97.
26 sites. (Map 21) Recorded from Shropshire, Denbighshire, Flintshire, Derbyshire and South Lancashire.

Licea parasitica (Zukal) G.W. Martin
On bark of living trees, in moist chamber; common.
First record 1976, last record 2011.   SD 90; SE 00; SJ 27–29, 35–39, 44–48, 54–59, 64–69, 74–79, 85, 87–89, 95–99; SK 09.   90 sites. (Map 22) Recorded in all neighbouring counties.

Licea pedicellata (H.C. Gilbert) H.C. Gilbert
On bark of living elder and willow trees, in moist chamber; uncommon.
First record 2003, last record 2011. SJ 27, 29, 46, 47, 59, 64, 75. Seven   sites.   (Map   23)   Recorded   from   Shropshire, Denbighshire and Flintshire.

Licea perexigua T.E. Brooks & H.W. Keller
On bark of living trees, in moist chamber; uncommon.
First record 2006, last record 2010. SJ 27–29, 59.   Five sites. Recorded from Flintshire and South Lancashire.

Licea pusilla Schrad.
On fallen pine logs and bark of living trees, in moist chamber; frequent.
First record 1965, last record 2010. SE 00; SJ 56, 57, 64, 68, 78, 99. Seven sites. (Map 24) Recorded in all neighbouring counties except Staffordshire.

Licea pygmaea (Meylan) Ing
On bark of living trees, in moist chamber; frequent.
First record 2003, last record 2010.   SJ 28, 29, 35, 38, 44–47, 54, 55, 58, 64, 67, 68, 74–77, 87, 89, 97–99.
25 sites. (Map 25) Recorded from Flintshire, South Lancashire and South West Yorkshire.

Licea sambucina D.W. Mitchell
On cyanobacteria on bark of living elder, oak and willow trees; uncommon.
Norton Priory, Crowden, Brabyn Park, Marple, 2008 and Alsager, 2010, all *B. Ing.* SJ 58, 75, 98; SK 09. Recorded from Flintshire. This species is usually found on elder, often with *L. microscopica*, but it has been found on several different trees in Britain and Europe.

Licea scintillans McHugh & D.W. Mitchell
On bark of living lime tree; rare.
Cheetham Park, Stalybridge, 2008, *B. Ing.* SJ 99. Not known in surrounding counties. This is the fifth record from the British Isles, being previously found in Ayrshire, Surrey, Sussex and Co. Kildare.

Licea scyphoides T.E. Brooks & H.W. Keller
On bark of living trees, in moist chamber; frequent, especially in the west.
First record 2007, last record 2010. SJ 29, 35, 46, 54, 65, 68, 75, 89, 97, 99. 14 sites. (Map 26) Recorded from Shropshire, Flintshire and Derbyshire. This species was earlier thought to be confined to Atlantic woodlands but is now not uncommon in city parks in several conurbations, including London.

Licea synsporos Nann.-Bremek.
On tips of moss leaves on bark of living poplar and sycamore; uncommon.
First record 2008, last record 2010. SJ 28, 59, 76, 98. Five sites. Recorded from Denbighshire and Flintshire.

Licea tenera E. Jahn
On bark of living apple tree, in moist chamber; rare.
Ness Gardens, 2010, *B. Ing.* SJ 37. Not known in surrounding vice-counties. This is only the seventh record from Britain of a much misunderstood species.

Licea testudinacea Nann.-Bremek.
On bark of living trees, in moist chamber; uncommon.
Moreton, 2006, Neston and Lymm Dam, 2010, *B. Ing.* SJ 27, 29, 68. Recorded from Derbyshire and South Lancashire.

Licea variabilis Schrad.
On fallen, decorticated conifer branches; common.
First record 1899, last record 2010. SE 00; SJ 28, 56, 57, 97. Six sites. (Map 27) Recorded in all neighbouring counties.

Family Dictydiaethaliaceae

Dictydiaethalium plumbeum (Schum.) Rostaf.
On fallen trunks, especially of beech or elm; frequent.
Marple Woods, 1956, *B. Hartham*; Arnfield, 1956, *S.S. Bates*; Styal Woods, 1960, *J.E. Milne*; Eaton Park, Chester and Lyme Park, 1972, *B. Ing.* SJ 46, 88, 98; SK 09. Recorded in all neighbouring counties.

Family Reticulariaceae

Lycogala epidendrum agg.
On fallen, rotten wood, especially in the spring; common.
This is an aggregate of two common species which were only separated in 1999. In the absence of herbarium specimens the older records cannot be assigned to *L. epidendrum* (L.) Fr. or *L. terrestre* Fr. so the records for the aggregate are given first.
First record 1911, last record 2010. SD 90; SE 00; SJ 27, 35-38, 44-48, 54-58, 64-69, 74-78, 87-89, 96-99; SK 09. 71 sites. (Map 28) Recorded in all neighbouring counties.

Lycogala epidendrum (L.) Fr.
On rotten wood; frequent.
This species has a scarlet plasmodium, grey or olivaceous spore mass and a rough surface to the dark aethalium, together with some microscopic differences.
First record 1997, last record 2008. SJ 28, 57, 78, 88, 99. Six sites. Recorded in Denbighshire, Flintshire, Derbyshire, Shropshire and South Lancashire.

Lycogala terrestre Fr.
On rotten wood; common.
This species may be recognised by the pink, never red, plasmodium, the pink spore mass and smooth surface.
First record 1911, last record 2010. SD 90; SE 00; SJ 27, 35-38, 44-48, 54-58, 64-69, 74-78, 87-89, 96-99; SK 09. 49 sites. (Map 29) Recorded in all neighbouring counties.

Lycogala exiguum Morgan
On fallen trunks, especially of beech; rare.
Delamere Forest, 1972, *B. Ing.* SJ 57. Recorded in Shropshire, Denbighshire, Flintshire and Derbyshire.

Lycogala flavofuscum Ehrenb.
On dead, standing trunks of broad-leaved trees; rare.
Eaton Park, Chester, 1979, *B. Ing. SJ* 46. Recorded from
Staffordshire, Shropshire and Denbighshire.

Reticularia jurana Meylan
Enteridium splendens Rostaf. var. *juranum* (Meylan) Härkönen
On fallen branches of broad-leaved trees in summer and
autumn; common.
Abbot's Moss, 1977, Rostherne Mere, 1985, Stanney Wood,
2008, and Lymm Dam, 2010, all *B. Ing.* SJ 37, 56, 68, 78.
Recorded in all surrounding counties except South West
Yorkshire.

Reticularia lobata Lister
Enteridium lobatum (Lister) M.L. Farr
Inside bark of pine stumps; uncommon.
Delamere Forest, 1910, *BMSF*; Storeton Hill, 1972, *B. Ing.* SJ 38,
57. Recorded in all neighbouring counties.

Reticularia lycoperdon Bull.
Enteridium lycoperdon (Bull.) M.L. Farr
On dead standing trunks, fallen logs and prepared timbers on
house doors and window frames; common.
First record 1909, last record 2010. SD 90; SJ 28, 36–38, 45–47,
54–58, 64–68, 75–78, 87, 88, 97–99; SK 09. 54 sites. (Map 30)
Recorded in all neighbouring counties.

Tubifera ferruginosa (Batsch) J.F. Gmel
Tubulifera arachnoidea Jacq.
On rotten conifer wood in late summer; common.
First record 1947, last record 2010. SE 00; SJ 36–38, 45, 46, 55–58, 64, 66, 67, 77–79, 87, 88, 97–99.
24 sites. (Map 31) Recorded in all neighbouring counties.

Order CRIBRARIALES
Family Cribrariaceae

Cribraria argillacea (Pers.) Pers.
On rotten conifer wood; common.
First record 1910, last record 2010. SE 00; SJ 28, 37, 38, 45, 46, 55–58, 64, 66, 67, 76–78, 86, 88, 97, 99; SK 09. 23 sites. (Map 32) Recorded in all neighbouring counties.

Cribraria aurantiaca Schrad.
On rotten conifer wood; common.
First record 1912, last record 2010. SJ 28, 37, 38, 45, 55–58, 64, 66, 67, 77, 78, 85, 88, 97. 17 sites. (Map 33) Recorded in all neighbouring counties.

Cribraria cancellata (Batsch) Nann.-Bremek.
On rotten conifer wood; common.
First record 1910, last record 2010. SJ 37, 38, 55–58, 64, 66, 67, 77, 85, 88, 97. 16 sites. (Map 34) Recorded in all neighbouring counties.

Cribraria dictydioides Cooke & Balf.
On litter in orchid house; rare – this is a tropical species which is occasionally imported with the plants.
Wilmslow, 1900, *A. Hodgkinson.* SJ 88. Not known in surrounding counties.

Cribraria intricata Schrad.
On rotten wood, especially of oak; rare.
Cotterill Clough, 1952, *S.S. Bates*; Delamere Forest, 1972, *B. Ing.* SJ 57, 88. Recorded from South Lancashire.

Cribraria persoonii Nann.-Bremek.
On rotten conifer wood; frequent.
Delamere Forest, 1972, *B. Ing.* SJ 57. Recorded in all neighbouring counties.

Cribraria rufa (Roth) Rostaf.
On rotten conifer wood; common.
First record 1910, last record 2010. SJ 28, 38, 55, 57, 66, 67, 77, 78, 87, 88, 97, 99. 19 sites. (Map 35) Recorded in all neighbouring counties.

Lindbladia tubulina Fr.
On conifer sawdust and stumps; rare.
Delamere Forest, 1972, *B. Ing.* SJ 57. Recorded from Derbyshire and Shropshire.

Order TRICHIALES
Family Dianemataceae

Calomyxa metallica (Berk.) Niewland
On bark of living elder, oak and elm, in moist chamber, and on litter; common.
First record 1943, last record 2011. SD 90; SJ 27–29, 35–39, 44–48, 54–59, 64–69, 74–79, 85–89, 96–99. 61 sites. (Map 36)
Recorded in all neighbouring counties.

Family Arcyriaceae

Arcyodes incarnata (Alb. & Schwein.) O.F. Cook
On soggy wood in dried-up ponds and alder woods; rare.
Alderley Edge, 1900, *W.H. Pepworth*; Stalybridge, 1956, *S.S. Bates*. SJ 87, 99. Recorded in all surrounding counties except Denbighshire and Flintshire.

Arcyria affinis Rostaf.
On fallen branches, especially of beech; uncommon.
Crows-i'th-Wood, 1952 and Arnfield, 1956, *S.S. Bates*; Styal Woods, 1965, *B. Ing*. SJ 88, 99. SK 09. Recorded from Staffordshire, Shropshire and Derbyshire.

Arcyria cinerea (Bull.) Pers.
On mossy rotten wood and bark of living trees, in moist chamber; common.
First record 1904, last record 2011. SJ 27, 28, 35–38, 44–47, 54–58, 64–69, 75, 77–79, 86–88, 96–99; SK 09. 49 sites. (Map 37)
Recorded in all neighbouring counties.

Arcyria denudata (L.) Wettst.
On rotten stumps and fallen trunks, especially of beech; common.
First record 1859, last record 2010. SJ 28, 37, 38, 45, 46, 54–58, 64–68, 76–78, 85–88, 97–99. 46 sites. (Map 38) Recorded in all neighbouring counties.

Arcyria ferruginea Sauter
On fallen logs and stumps, especially in winter; frequent.
First record 1909, last record 1980. SJ 28, 37, 38, 56, 57, 78, 86–88, 98, 99; SK 09. 16 sites. (Map 39) Recorded in all neighbouring counties.

Arcyria incarnata (Pers.) Pers.
On fallen branches, especially of oak and often on the broken ends; common.
First record 1859, last record 2010. SJ 37, 38, 46, 55–58, 64–68, 77, 78, 86, 87, 97, 99. 20 sites. (Map 40) Recorded in all neighbouring counties.

Arcyria minuta Buchet
On fallen bark and small branches; uncommon.
Arnfield, 1956, *S.S. Bates*; Eaton Park, Chester, 1988, *B. Ing*; Dibbinsdale, 2001 and Cotterill Clough, 2003, *NWFG*. SJ 38, 46, 88; SK 09. Recorded in all surrounding counties except Staffordshire and South West Yorkshire.

Arcyria obvelata (Oeder) Onsberg
On logs and stumps, often drier than usual for myxomycetes; common.
First record 1895, last record 2010. SJ 38, 55–57, 65–68, 76–78, 86, 88, 97–99; SK 09. 21 sites. (Map 41) Recorded in all neighbouring counties.

Arcyria oerstedtii Rostaf.
On stumps and fallen trunks of beech and conifers; uncommon.
Arnfield, 1956, *S.S. Bates*; Farnworth, Widnes, 1957, *B. Hartham*; Rostherne Mere, 1964, *J.E. Milne*; Delamere Forest, 1984, *B. Ing*. SJ 57, 58, 78; SK 09. Recorded in all neighbouring counties.

Arcyria pomiformis (Leers) Rostaf.
On fallen trunks and bark of living trees; common.
First record 1941, last record 2011. SJ 27, 28, 35–39, 44–47, 54–58, 64–69, 74–79, 85–87, 89, 96–99; SK 09. 57 sites. (Map 42) Recorded in all neighbouring counties.

Perichaena chrysosperma (Currey) Lister
On the bark of living trees, in moist chamber; common.
First record 2007, last record 2011. SD 90; SJ 27–29, 35–38, 44–48, 54–59, 64–69, 74–79, 85, 87–89, 96–99; SK 09. 57 sites. (Map 43) Recorded in all neighbouring counties.

Perichaena corticalis (Batsch) Rostaf.
Under bark of fallen trees, especially ash; common.
First record 1872, last record 2010. SJ 28, 38, 46, 68, 78, 85–88, 99; SK 09. 19 sites. (Map 44) Recorded in all neighbouring counties.

Perichaena depressa Libert
Under bark of fallen ash trunks; frequent.
First record 1950, last record 2011. SJ 38, 59, 88, 99; SK 09. Six sites. (Map 45) Recorded in all neighbouring counties.

Perichaena vermicularis (Schwein.) Rostaf.
On leaf litter, especially of sycamore, and rarely on bark of living trees; uncommon.
First record 1944, last record 2010.   SJ 29, 38, 46, 86; SK 09.
Five sites.   Recorded in all surrounding counties except Staffordshire and Denbighshire.

Family Trichiaceae

Hemitrichia calyculata (Speg.) M.L. Farr
On rotten logs, especially of beech; frequent.
First record 1911, last record 1974.   SJ 37, 78, 88, 98; SK 09.
Eight sites. (Map 46) Recorded in all neighbouring counties.

Hemitrichia leiotricha (Lister) G. Lister
On dead stems in deep heather; rare.
Thurstaston Common, 1985, *B. Ing.*   SJ 28.   Recorded from Shropshire, Denbighshire, Derbyshire and South West Yorkshire.

Hemitrichia pardina (Minakata) Ing
On bark of living poplar, sycamore and willow trees; uncommon.
Crewe and Woodbank Park, 2007, Burtonwood, 2008 and Lymm Dam, 2010, all *B. Ing.* SJ 59, 68, 75, 99.   Recorded from Denbighshire and South Lancashire.

Metatrichia floriformis (Schwein.) Nann.-Bremek.
On rotten trunks and branches; common.
First record 1952, last record 2011.   SD 90; SJ 27, 28, 35–38, 45–47, 54–58, 64–69, 74, 78–79, 87, 88, 96, 97, 99; SK 09. 42 sites.
(Map 47)   Recorded in all neighbouring counties.

Metatrichia vesparium (Batsch) Nann.-Bremek.
On old rotten trunks, especially of elm and beech; uncommon in the northern half of England, common in the south.
First record 1941, last record 1970. SJ 37, 78, 86, 88, 99. Seven sites. (Map 48) Recorded in all surrounding counties except Denbighshire.

Oligonema schweinitzii (Berk.) G.W. Martin
On sticks in dried-up ponds or on spent tan; rare.
Ditton Junction and Bidston Hill, 1963, *J.T. Palmer*; Arrowe Park, 1972, *B. Ing.* SJ 28, 48.   Recorded from Staffordshire, Shropshire, Flintshire and South Lancashire.

Prototrichia metallica (Berk.) Massee
On twiggy litter in woodland, especially in winter; uncommon.
Buglawton, 1944, *W.D. Graddon*; Eccleston, 1972, *B. Ing.* SJ 46, 86. Recorded in all surrounding counties except South Lancashire.

Trichia affinis de Bary
On very rotten wood, usually on mosses; common.
First record 1827, last record 2010.   SJ 28, 36–38, 45–47, 54–58, 64–69, 75–79, 86–88, 96, 97, 99. 38 sites. (Map 49) Recorded in all neighbouring counties.

Trichia botrytis (J.F. Gmel.) Pers. var. **botrytis**
On rotten trunks and branches, especially of oak and conifers; common.
First record 1895, last record 2010. SE 00; SJ 28, 36–38, 45–47, 54–58, 64–69, 75–78, 86–88, 97, 98; SK 09. 40 sites. (Map 50) Recorded in all neighbouring counties.

Trichia botrytis var. **cerifera** G. Lister
On fallen branches; rare.
Cotterill Clough, 1963, *J.E. Milne*. SJ 88. Recorded from Derbyshire.

Trichia contorta (Ditmar) Rostaf. var. **contorta**
On fallen branches and trunks; frequent.
First record 1960, last record 2010. SJ 46, 77, 78, 88. Six sites. (Map 51)   Recorded in all neighbouring counties.

Trichia contorta var. **inconspicua** (Rostaf.) Lister
On small twigs and leaf litter; frequent.
Great Moreton Hall, 1943, *W.D. Graddon*; Swineshaw, 1953, *S.S. Bates*.   SJ 85; SK 09.   Recorded from Staffordshire, Denbighshire, Derbyshire and South West Yorkshire.

Trichia decipiens (Pers.) T. Macbr.
On rotten trunks and branches; common.
These records probably include the recently separated *T. meylanii* Ing, which was previously known as the var. *olivacea* Meylan of *T. decipiens*.   Herbarium specimens have not yet been re-examined to see whether the new species was present in Cheshire prior to the record below.
First record 1912, last record 2010.   SJ 27, 28, 35–38, 45–47, 54–59, 64–69, 75–78, 85–88, 97, 98; SK 09. 37 sites.   (Map 52) Recorded in all neighbouring counties.

Trichia flavicoma (Lister) Ing
In leaf litter; rare.
Arnfield, 1956, *S.S. Bates*.   SK 09.   Recorded from Shropshire, Derbyshire and South West Yorkshire.

Trichia meylanii Ing
On rotten trunks and branches; probably common.
Eastham Wood, 2011, *B. Ing.* SJ 38.   Recorded from
Denbighshire, Flintshire, Derbyshire and Shropshire. This
species is separated from *T. decipiens* by the circumscissile
dehiscence of the sporocarp and the warted, rather than finely
reticulated spores.

Trichia munda (Lister) Meylan
On mosses on the bark of living trees; uncommon.
Buglawton, 1942, *W.D. Graddon*; Delamere Forest, 1979,
Grosvenor Park, Chester, 2009 and Nantwich, 2010, *B. Ing.* SJ
46, 57, 65, 86. Recorded in all surrounding counties except
Flintshire and South Lancashire.

Trichia persimilis P. Karst.
On hard rotten wood; common.
First record 1902, last record 2010.   SJ 28, 35–38, 45–47, 54–58,
64–69, 75–78, 86–88, 96–99; SK 09. 38 sites. (Map 53) Recorded
in all neighbouring counties.

Trichia scabra Rostaf.
On large, old, rotten trunks, especially of elm or beech;
frequent.
First record 1910, last record 2006. SJ 37, 38, 46, 57, 67, 76, 79,
85, 86, 88; SK 09. 13 sites. (Map 54) Recorded in all
neighbouring counties.

Trichia sordida Johann.
On living heather stems near melting snow, moorland; rare.
Woodhead Pass, 2010, *B. Ing.* SE 00. This low-alpine species was added to the British list in 2000 and is known from several localities in the Pennines in Derbyshire and South West Yorkshire, a few miles from the Woodhead site. The species is found mainly in February and is probably more widespread than the records suggest.

Trichia varia (Pers.) Pers.
On soggy, rotten wood; common.
First record 1828, last record 2011. SD 90; SJ 27, 28, 35–38, 44–48, 54–58, 64–69, 74–78, 86–89, 96, 97, 99; SK 09. 62 sites. (Map 55) Recorded in all neighbouring counties.

Trichia verrucosa Berk.
On damp, rotten conifer trunks; rare.
Delamere Forest, 1972, *B. Ing.* SJ 57. Recorded in all surrounding counties except Staffordshire and South Lancashire.

Order STEMONITALES
Family Stemonitidaceae

Amaurochaete atra (Alb. & Schwein.) Rostaf.
On newly felled conifer trunks; uncommon.
First record 1910, last record 1977. SJ 56, 57, 67, 78, 88, 98. Six sites. (Map 56) Recorded in all surrounding counties except Derbyshire.

Amaurochaete tubulina (Alb. & Schwein.) T. Macbr.
On worked timber in newly constructed house; rare.
Bromborough, 1964, *G. Wimpey*.   SJ 38.   Recorded from
Staffordshire.

Brefeldia maxima (Fr.) Rostaf.
On stumps; uncommon.
This is the largest myxomycete known, the fruit bodies may
be in excess of one square metre in area.
First record 1959; last record 2001.   SJ 46, 57, 88, 99.   Six sites.
Recorded in all neighbouring counties.

Collaria elegans (Racib.) Dhillon & Nann.-Bremek.
On decorticated conifer sticks on the forest floor; frequent.
Delamere Forest, 1974, *B. Ing*; Styal Woods, 2001, *NWFG*.   SJ
57, 88. Recorded in all surrounding counties except South
West Yorkshire.

Colloderma oculatum (Lippert) G. Lister
On mosses and lichens on the bark of living trees, in moist
chamber; uncommon.
First record 1976, last record 2010. SJ 28, 46, 56, 86.   Five sites.
Recorded in all surrounding counties except Staffordshire.

Comatricha alta Preuss
On old logs, especially on the cut end; uncommon.
Congleton, 1944, *W.D. Graddon*; New Brighton, 1999, *A. Ryan*.
SJ 39, 86. Recorded in all surrounding counties except
Staffordshire and Shropshire.

Comatricha ellae Härkönen
On bark of living oak in moist chamber; rare.
Delamere Forest, 1993, *D. Wrigley–Basanta.* SJ 57. Recorded
from South West Yorkshire.

Comatricha laxa Rostaf.
On dead wood, and bark of living trees; frequent.
First record 2007, last record 2010. SJ 37, 45, 47, 55, 57, 66, 67,
74. Eight sites. (Map 57) Recorded from all surrounding
counties except Flintshire.

Comatricha nigra (Pers.) Schröt.
On dead wood of all kinds, occasionally on bark of living
trees; common.
First record 1899, last record 2011. SD 90; SE 00; SJ 27, 28,
35–38, 44–48, 54–59, 64–69, 74–79, 85–89, 96–99. 66 sites. (Map
58) Recorded in all neighbouring counties.

Comatricha pulchella (C. Bab.) Rostaf.
On leaf litter, especially of holly, and bases of old fern fronds;
common.
First record 1899, last record 2011. SJ 37, 38, 45–47,
55–57, 64, 68, 78, 86, 87, 98, 99. 21 sites. (Map 59) Recorded in
all neighbouring counties.

Comatricha rigidireta Nann.-Bremek.
On acid bark of living trees; uncommon.
First record 2007, last record 2010. SJ 37, 44, 46, 47, 55, 57, 68,
88. Eight sites. (Map 60) Recorded from Shropshire and
Denbighshire.

Comatricha tenerrima (M.A. Curtis) G. Lister
On dead herbaceous stems in damp sites; uncommon.
First record 1941, last record 2010.   SD 90; SJ 57, 64, 65, 68, 86.
Seven sites. (Map 61) Recorded in all surrounding counties.

Enerthenema papillatum (Pers.) Rostaf.
On the bark of living trees, in moist chamber, and on fallen branches; common.
First record 1941, last record 2011.   SD 90; SE 00; SJ 28, 36–38, 44–47, 54–58, 64–69, 74–78, 85–89, 96–99. 42 sites.   (Map 62) Recorded in all neighbouring counties.

Lamproderma arcyrioides (Sommerf.) Rostaf.
On leaf litter, especially of ivy; frequent.
Lower Heath Wood, Storeton, 1911, *J.W. Ellis*. SJ 38.
Recorded in all neighbouring counties.

Lamproderma columbinum (Pers.) Rostaf.
On mossy trunks and stumps; uncommon, except in the west of Britain.
Eastham Wood, 1912, *J.W. Ellis*; Broadbottom, 1959, *S.S. Bates*; Cotterill Clough, 1964, *J.E. Milne*; Abbot's Moss, 1977 and Hatchmere, Delamere Forest, 1977, *B. Ing*. SJ 38, 56, 57, 88, 99. Recorded in all surrounding counties except Staffordshire.

Lamproderma echinulatum (Berk.) Rostaf.
On stumps; rare.
Marple Wood, 1956 and Broadbottom Wood, 1959, *S.S. Bates*. SJ 98, 99. Recorded from Derbyshire and South West Yorkshire.

Lamproderma scintillans (Berk. & Broome) Morgan
On leaf litter, especially of holly, and on wet, rotten fern fronds; common.
First record 1942, last record 2011. SJ 28, 37, 38, 45–47, 55–57, 67, 68, 76, 77, 86–88, 98, 99. 18 sites. (Map 63) Recorded in all neighbouring counties.

Macbrideola cornea (L. Lister & Cran) Alexop.
On mossy bark of living poplar tree, in moist chamber; common in the west.
Woodbank Park, 2007, *B. Ing*.   SJ 99.   Recorded from all neighbouring counties.

Paradiacheopsis cribrata Nann.-Bremek.
On the bark of living oak trees; frequent.
First record 1997, last record 2010.   SJ 47, 57, 58, 67, 75, 88. Seven sites. (Map 64) Recorded in all surrounding counties except Staffordshire.

Paradiacheopsis fimbriata (G. Lister & Cran) Hertel
On acid bark of living trees; common, especially in urban areas.
First record 1924, last record 2011.   SD 90; SE 00; SJ 28, 35–38, 44–48, 54–59, 64–68, 74–79, 85, 87–89, 96–99. 58 sites. (Map 65) Recorded in all neighbouring counties.

Paradiacheopsis microcarpa (Meylan) D.W. Mitchell
On acid bark of living pine tree, in moist chamber; uncommon.
Delamere Forest, 1993, *D. Wrigley-Basanta*. SJ 57. Recorded from South Lancashire.

Paradiacheopsis rigida (Brandza) Nann.-Bremek.
On bark of living rhododendron tree, in moist chamber; rare.
Daresbury, 2010, *B. Ing.* SJ 58. Recorded from Denbighshire and South Lancashire.

Paradiacheopsis solitaria (Nann.-Bremek.) Nann.-Bremek.
On bark of living trees, in moist chamber; common.
First record 1979, last record 2011. SJ 36–38, 44–47, 54–58, 64–69, 75–78, 86, 87,89, 96–99.
36 sites. (Map 66) Recorded in all neighbouring counties.

Stemonitis axifera (Bull.) T. Macbr.
On fallen branches of broad-leaved trees; common.
First record 1895, last record 2010. SD 90; SJ 37, 46, 55–57, 68, 78, 86–88, 99; SK 09. 15 sites. (Map 67) Recorded in all neighbouring counties.

Stemonitis flavogenita E. Jahn
On rotten trunks and branches; common.
First record 1895, last record 1985. SJ 28, 56, 57, 78, 88, 99.
Nine sites. (Map 68) Recorded in all neighbouring counties.

Stemonitis fusca Roth
On stumps and rotten trunks and branches; common.
First record 1895, last record 2011. SD 90; SJ 28, 37, 38, 44-47, 54–58, 64–68, 75–78, 86–88, 96–99.
40 sites. (Map 69) Recorded in all neighbouring counties.

Stemonitis herbatica Peck
On leaf litter on the forest floor; uncommon.
Delamere Forest, 1980, *B. Ing.* SJ 57. Recorded in all surrounding counties except Staffordshire and South Lancashire.

Stemonitis nigrescens Rex
On fallen branches; rare, more usually on bark of living trees.
Stalybridge, 1956, *S.S. Bates*; Sun Bank Wood, 2002, *P.R. Smith*.
SJ 78, 99.   Recorded from South Lancashire.

Stemonitopsis hyperopta (Meylan) Nann.-Bremek.
On very rotten conifer wood; frequent.
First record 1971, last record 2010.   SE 00; SJ 28, 56, 57, 64, 97.
Six sites. Recorded in all surrounding counties.

Stemonitopsis typhina (F.H.Wigg.) Nann.-Bremek.
On soggy rotten wood; common.
First record 1951, last record 2011.   SD 90; SJ 27, 28, 35, 37, 38,
44–47, 54–58, 64–68, 75–78, 87, 88, 96–99; SK 09.   38 sites.
(Map 70) Recorded in all neighbouring counties.

Symphytocarpus amaurochaetoides Nann.-Bremek.
On stumps; uncommon.
Radnor Hill, 1941, *W.D. Graddon*; Chester Zoo, 2006, *B. Ing*. SJ
47, 86. Recorded from Staffordshire, Shropshire, Denbighshire
and South West Yorkshire.

Symphytocarpus flaccidus (Lister) Ing & Nann.-Bremek.
On dead standing pine trees; frequent.
Thurstaston Common, 1971, Delamere Forest, 1972 and
Abbot's Moss, 1978, all *B. Ing*.   SJ 28, 56, 57. Recorded in all
surrounding counties.

Order PHYSARALES
Family Physaraceae

Badhamia affinis Rostaf.
On mossy bark of living trees; frequent in the west.
First record 1957, last record 2010. SJ 29, 48, 66, 68, 86; SK 09.
Seven sites. (Map 71) Recorded from Denbighshire, Flintshire and South Lancashire.

Badhamia capsulifera (Bull.) Berk.
On bark of living branches of trees; uncommon.
Bramhall Park, 1961, *J.E. Milne*. SJ 88. Recorded from Shropshire, Denbighshire and Flintshire.

Badhamia foliicola Lister
On living grasses, litter and the bark of living trees, in moist chamber; frequent.
Styal Wood, 1965, Delamere Forest, 1976, *B. Ing*; Chester University campus, 1979, *N. Hughes*. SJ 46, 57, 88. Recorded from Staffordshire, Shropshire, Flintshire, South Lancashire and Derbyshire.

Badhamia lilacina (Fr.) Rostaf.
On the surface of living *Sphagnum* and other plants in bogs; uncommon.
Thurstaston Common, 1971, Abbot's Moss, 1977, Hatchmere, Delamere Forest, 1977 and Wybunbury Moss, 1978, all *B. Ing*. SJ 28, 56, 57, 65. Recorded from Shropshire, Denbighshire, Flintshire and South West Yorkshire.

Badhamia macrocarpa (Ces.) Rostaf.
On fallen bark and branches; uncommon.
First record 1950, last record 1976. SJ 68, 88, 99; SK 09. Five sites. Recorded in all surrounding counties except Denbighshire and Flintshire.

Badhamia panicea (Fr.) Rostaf.
On fallen trunks, especially of beech, rarely on bark in moist chamber; common.
First record 1896, last record 2010. SJ 37, 47, 55–57, 65, 66, 68, 78, 87, 88, 98, 99; SK 09. 20 sites. (Map 72)   Recorded in all neighbouring counties.

Badhamia utricularis (Bull.) Berk.
On fungi such on *Stereum hirsutum* and *Phlebia radiata* on fallen trunks; common.
First record 1899, last record 2011. SJ 35, 37, 38, 45–47, 55–57, 64–69, 77, 78, 85–88, 98, 99. 33 sites. (Map 73) Recorded in all neighbouring counties.

Craterium aureum (Schumach.) Rostaf.
In leaf litter, especially of beech; uncommon.
Buglawton, 1943, *W.D. Graddon*; Delamere Forest, 1985, *B. Ing*. SJ 57, 86. Recorded in all surrounding counties except South West Yorkshire.

Craterium leucocephalum (Pers.) Ditmar
In leaf litter and mosses; frequent.
First record 1912, last record 1965. SJ 29, 85, 86, 88, 99; SK 09. Six sites. (Map 74) Recorded in all neighbouring counties.

Craterium minutum (Leers) Fr.
In leaf litter and on living herbaceous stems; common.
First record 1910, last record 2011. SJ 28, 36–38, 44–47, 54–59, 64–68, 75–78, 86–88, 97–99; SK 09. 45 sites. (Map 75) Recorded in all neighbouring counties.

Craterium muscorum Ing
On mosses on wet rocks in ravines and very wet woodland; rare.
Cotterill Clough, 1854, *W. Wilson*, (Smith & Rea, 1905.) SJ 88. Recorded from Shropshire, otherwise the nearest sites are in Snowdonia, but probably extinct in Cheshire. The published references cite 'Clough' in Cheshire, but Wilson collected intensively at Cotterill Clough from 1827 onwards, including several species of myxomycetes over many years.

Fuligo cinerea (Schwein.) Morgan
On manure heaps; probably extinct in Britain.
Carrington Moss, 1904, *H. Murray*. SJ 79. Recorded from Staffordshire and Denbighshire.

Fuligo rufa Pers.
On exposed, very rotten stumps and old straw; uncommon.
Styal Wood, 1964, *S.S. Bates*; Delamere Forest, 1972 and Dunham Massey park, 1983, *B. Ing*. SJ 57, 78, 88. Recorded in all surrounding counties except Denbighshire.

Fuligo septica (L.) F.H. Wigg.
On rotten trunks and branches, on mosses near wood and on tan bark; common.
First record 1910, last record 2010.   SD 90; SJ 28, 36–38, 44–48, 54–58, 64–68, 75–78, 86–88, 96–99. 50 sites. (Map 76)   Recorded in all neighbouring counties.

Leocarpus fragilis (Dicks.) Rostaf.
In leaf litter and climbing up herbaceous stems; common.
First record 1899, last record 2010.   SE 00; SJ 28, 29, 37, 38, 44–48, 54–57, 64–68, 75–78, 86–88, 97–99; SK 09.   41 sites. (Map 77) Recorded in all neighbouring counties.

Physarum album (Bull.) Chevall.
Physarum nutans Pers.
On dead wood on the forest floor; common.
First record 1895, last record 2010.   SD 90; SJ 27, 28, 35–38, 44–47, 54–58, 64–69, 75–78, 86–88, 96, 97, 99; SK 09. 53 sites.   (Map 78)   Recorded in all neighbouring counties.

Physarum auriscalpium Cooke
On bark of living trees; uncommon.
First record 2008, last record 2010.   SJ 29, 58, 66, 68, 69, 77. Six sites. (Map 79) Recorded from South Lancashire.

Physarum bitectum G. Lister
On bramble stems and leaf litter; uncommon.
First record 1912, last record 1988.   SJ 28, 46, 56, 77, 78, 85, 86, 88. Nine sites. (Map 80)   Recorded in all surrounding counties except South Lancashire.

Physarum bivalve Pers.
In leaf litter; common.
First record 1941, last record 2010. SJ 28, 38, 45, 46, 55–57, 59, 65, 66, 68, 78, 86; SK 09. 21 sites. (Map 81) Recorded in all neighbouring counties.

Physarum brunneolum (Phill.) Massee
On leaf litter; rare.
Congleton, 1943, *W.D. Graddon*. SJ 86. This species is rarely found north of the Mediterranean and has been found only three times in the British Isles, after hot summers and mild winters; the last occasion was in 1976, in Kent, and the previous time was in Co. Down in 1911.

Physarum cinereum (Batsch) Pers.
In leaf litter and on grass in lawns; common.
First record 1960, last record 2010. SJ 36, 46, 65, 68, 78, 87, 88. Nine sites. (Map 82) Recorded in all neighbouring counties.

Physarum compressum Alb. & Schwein.
On straw, herbaceous waste and compost heaps; common.
First record 1910, last record 1987. SJ 38, 57, 78, 88, 98, 99; SK 09. 12 sites. (Map 83) Recorded in all neighbouring counties.

Physarum contextum (Pers.) Pers.
On leaf litter and mosses; uncommon.
Little Budworth Common, 1985, *B. Ing*. SJ 56. Recorded in all surrounding counties except Flintshire.

Physarum crateriforme Petch
On bark of living trees; uncommon.
Poynton Park, 2007, Helsby Quarry and Shakerley Mere, 2010 and Chester Zoo, 2011, *B. Ing.* SJ 47, 77, 98. Recorded from Shropshire and Denbighshire.

Physarum didermoides (Pers.) Rostaf.
On straw heaps and grass litter; rare and declining as its main habitat disappears.
Eastham Rake Plantation, 1912, *J.W. Ellis*; (Item) Chester Zoo, 2011, *B. Ing.* SJ 37, 47. Recorded from Flintshire and Cheshire.

Physarum leucophaeum Fr.
On fallen branches and sticks; common.
First record 1951, last record 2010. SJ 28, 37, 38, 46, 57, 58, 68, 76, 78, 87, 98, 99; SK 09. 14 sites. (Map 84) Recorded in all neighbouring counties.

Physarum licheniforme (Schwein.) Lado
Physarum lividum Rostaf.
On straw heaps and grass litter; rare and declining.
Broadbottom Wood, 1958, *S.S. Bates.* SJ 99. Recorded from Derbyshire.

Physarum oblatum T. Macbr.
On mossy bark of living trees, in moist chamber; uncommon.
University of Chester campus, 1979, *N. Hughes*; Little Budworth Country Park, 2010, *B. Ing.* SJ 36, 56. Not known in surrounding counties.

Physarum obscurum (Lister) Ing
In leaf litter; rare.
Delamere Forest, 1985, *B. Ing.* SJ 57. Recorded from Staffordshire.

Physarum pusillum (Berk. & M.A. Curt.) G. Lister
On grass litter, including marram in sand dunes, rarely on bark; frequent.
First record 1941, last record 2010. SJ 28, 29, 39, 86; SK 09. Recorded in all surrounding counties except Staffordshire and South West Yorkshire.

Physarum robustum (Lister) Nann.-Bremek.
On rotten wood; frequent.
Rostherne, 1903, *H. Murray*; Stalybridge, 1954, *S.S. Bates*; Alderley Edge, 1981, *B. Ing.* SJ 78, 87, 99. Recorded in all surrounding counties except Denbighshire.

Physarum virescens Ditmar
On terrestrial feather-mosses in damp woodland; frequent in the west.
Abbot's Moss, 1977, *B. Ing.* SJ 56. Recorded from Staffordshire, Shropshire, Denbighshire and South West Yorkshire.

Physarum viride (Bull.) Pers.
On twigs and small branches, especially conifer brashings; frequent.
First record 1912, last record 2010. SJ 28, 37, 38, 56, 57, 66, 78, 88, 97, 99. 12 sites. (Map 85) Recorded in all neighbouring counties.

Family Didymiaceae

Diachea leucopodia (Bull.) Rostaf.
In leaf litter and on living herbaceous stems; frequent.
First record 1942, last record 1985.   SJ 46, 55–57, 78, 86, 99.
Eight sites.   (Map 86)   Recorded in all neighbouring counties.

Diderma chondrioderma (de Bary & Rostaf.) G. Lister
On mossy bark of living trees, in moist chamber; uncommon.
First record 1982, last record, 2010.   SD 90;   SE 00; SJ 36, 39, 47, 55, 58, 78, 79, 89. 10 sites. (Map 87) Recorded from Shropshire, Denbighshire, Flintshire, South Lancashire and Derbyshire.

Diderma deplanatum Fr.
On terrestrial mosses and leaf litter; uncommon.
Buglawton, 1943, *W.D. Graddon*; Delamere Forest, 1962, *J.E. Milne*.   SJ 57, 86. Recorded in all surrounding counties except South Lancashire.

Diderma donkii Nann-Bremek.
In leaf litter, especially of beech; rare.
Delamere Forest, 1984, *B. Ing*. SJ 57. Recorded from Shropshire and Derbyshire.

Diderma effusum (Schwein.) Morgan
In leaf litter and, occasionally, on bark of living trees, as a casual; frequent.
First record 1963, last record 2010.   SJ 46, 67, 78, 87, 88, 98, 99.
Seven sites. (Map 88) Recorded in all neighbouring counties.

Diderma globosum Pers.
In leaf litter and on living plants in damp sites; frequent.
Woodley, 1953, *J.T. Palmer*; Cotterill Clough, 1974, *B. Hartham*;
Little Budworth Common and Rostherne Mere, 1985, *B. Ing*.
SJ 56, 78, 88, 99. Recorded in all surrounding counties except
Staffordshire and South West Yorkshire.

Diderma hemisphaericum (Bull.) Hornem.
In leaf litter in damp sites, or as a casual on the bark of living
trees; frequent.
First record 1988, last record 2010. SD 90; SJ 39, 45, 46, 64, 65,
68, 76, 96, 98. 11 sites. (Map 89) Recorded in all
neighbouring counties.

Diderma montanum (Meyl.) Meyl.
In leaf litter in damp sites; rare.
Abbot's Moss, 1980, *B. Ing*. SJ 56. Recorded from Shropshire
and Derbyshire.

Diderma radiatum (L.) Morgan
On twiggy litter and bramble stems; rare.
Great Wood, Broadbottom, 1959, *J.T. Palmer*. SJ 99. Recorded
from Staffordshire, Derbyshire and South West Yorkshire.

Diderma simplex (Schröt.) G. Lister
In birch litter on acid soil; rare.
Delamere Forest, 1972, *B. Ing*. SJ 57. Recorded from South
West Yorkshire.

Diderma spumarioides (Fr.) Fr.
On mosses in sandy heaths and dunes; uncommon.
Rudheath, 2000 and Leasowe Common, 2006, *B. Ing.* SJ 29, 77.
Recorded in all surrounding counties except South West Yorkshire.

Diderma testaceum (Schrad.) Pers.
In leaf litter; rare.
Alderley Edge, 1953, *S.S. Bates*; Rostherne Mere, 1985, *B. Ing.*
SJ 78, 87. Recorded from Shropshire and Flintshire.

Diderma umbilicatum Pers.
On small branches and bramble stems; uncommon.
Priesty Fields, 1941, Congleton, 1943 and Great Moreton Hall, 1944, *W.D. Graddon*; Swineshaw Lodge, Stalybridge, 1960, *S.S. Bates.* SJ 85, 86. SK 09. Recorded in all surrounding counties except Denbighshire.

Didymium anellus Morgan
In leaf litter; uncommon, a southern species in Britain.
Cotterill Clough, 1950, *B. Hartham*; Styal, 1956 and Botham's Hall Wood, Broadbottom, 1960, *S.S. Bates.* SJ 88, 99.
Recorded from Shropshire, Flintshire and Derbyshire.

Didymium bahiense Gottsberger
In leaf litter, decaying herbaceous material and on rabbit dung; common.
First record 1910, last record 2010. SD 90; SJ 28, 46, 55–57, 64, 67, 68, 98. 11 sites. (Map 90) Recorded in all neighbouring counties.

Didymium clavus (Alb. & Schwein.) Rabenh.
In leaf litter, especially of oak; common.
First record 1943, last record 2010.   SJ 46, 56, 57, 86.   Six sites.
(Map 91)   Recorded in all neighbouring counties.

Didymium crustaceum Fr.
In leaf litter, especially in hedge bottoms; rare.
Delamere Forest, 1974, *B. Ing.*   SJ 57.   Recorded from
Denbighshire, Flintshire and South Lancashire.

Didymium difforme (Pers.) Gray
On leaf litter and herbaceous waste, occasionally on bark of
living trees; common.
First record 1904, last record 2011. SD 90; SE 00; SJ 27–29,
35–39, 44–48, 54–59, 64–69, 74–79, 85–87, 89, 96–99; SK 09. 67
sites. (Map 92) Recorded in all neighbouring counties.

Didymium ilicinum Ing
Usually on holly leaf litter but here also on bark of living lime
tree; frequent.
First record 1976, last record 2011. SJ 27, 37, 38, 45–47, 55, 56,
64, 67, 68, 76, 77, 87, 98, 99. 19 sites. (Map 93) Recorded from
Denbighshire, Derbyshire and South West Yorkshire.   This is
a segregate from the *D. squamulosum* complex, characterised
by the intricately ridged and flaky deposits on the sporangium
wall.

Didymium megalosporum Berk. & M.A. Curt.
In leaf litter and on rabbit dung, in moist chamber; frequent.
Staley, 1954, *S.S. Bates*; Bramhall Park, 1960, *J.E. Milne*;
Peckforton, 1977, *B. Ing.* SJ 55, 88, 99. Recorded from
Staffordshire, Flintshire, South Lancashire and South West
Yorkshire.

Didymium melanospermum (Pers.) T. Macbr.
On conifer and gorse litter; common.
First record 1910, last record 2010. SE 00; SJ 28, 37, 38, 46, 56, 57, 64, 87, 88, 97–99. 18 sites. (Map 94) Recorded in all neighbouring counties.

Didymium minus (Lister) Morgan
On dead herbaceous stems, especially of rosebay willow-herb; frequent.
Styal Woods, 1963, *J.E. Milne*; Delamere Forest, 1972, Eaton Park, Chester, 1988 and Marple, 2008, all *B. Ing*. SJ 46, 57, 88, 98. Recorded in all neighbouring counties.

Didymium nigripes (Link) Fr.
In leaf litter, especially of holly; common.
First record 1859, last record 2011. SJ 27, 28, 37, 38, 45-47, 55-59, 67, 68, 76, 77, 86-88, 98, 99. 26 sites. (Map 95) Recorded in all neighbouring counties.

Didymium serpula Fr.
In leaf litter, especially of oak; frequent in the south, rarer northwards.
Cotterill Clough, 1942, *S.S. Bates*; Rostherne Mere, 1964, *J.E. Milne*. SJ 78, 88. Recorded from Shropshire, Denbighshire, Flintshire, South Lancashire and South West Yorkshire.

Didymium squamulosum (Alb. & Schwein.) Fr. agg.
In leaf litter of all kinds; common.
This is a complex of asexual species and the older records from Cheshire may contain *D. ilicinum* and the next species.
First record 1859, last record 2011. SD 90; SE 00; SJ 27-29, 35-39, 44-48, 54-59, 64-69, 74-78, 86-89, 96-99; SK 09. 66 sites. (Map 96) Recorded in all neighbouring counties.

Didymium tussilaginis (Berk. & Broome) Massee
In leaf litter; rare, or at least rarely reported – part of the *D. squamulosum* complex.
Chester, 1872, *T. Brittain*. SJ 46. Originally described from Chester; not known in surrounding counties.

Didymium verrucosporum A.L. Welden
In grass litter; rare.
University of Chester campus, 1976, *B. Ing*. SJ 36. Recorded from Staffordshire, Flintshire, South Lancashire and South West Yorkshire.

Lepidoderma chailletii Rostaf.
In birch litter on acid soils; rare.
Delamere Forest, 1972, *B. Ing*. SJ 57. Recorded from Flintshire.

Mucilago crustacea F.H. Wigg.
Encrusting grass stems in grassland, roadsides and lawns; common.
First record 1859, last record 2010. SJ 28, 29, 36-38, 45-47, 54, 56-58, 64, 67, 68, 77, 78, 86-88. 25 sites. (Map 97) Recorded in all neighbouring counties.

Figure 1: *Licea erddigensis* – an uncommon species on bark, described from Erddig Park, Wrexham (picture by A. Orange).

Figure 2: *Licea eleanorae* – an uncommon species on bark, described in honour of the author's wife (picture by M. Lenne).

Figure 3: *Enerthenema papillatum* -- a common species on bark and fallen branches; the genus was originally described by a Wrexham naturalist (picture by P. Bruskern).

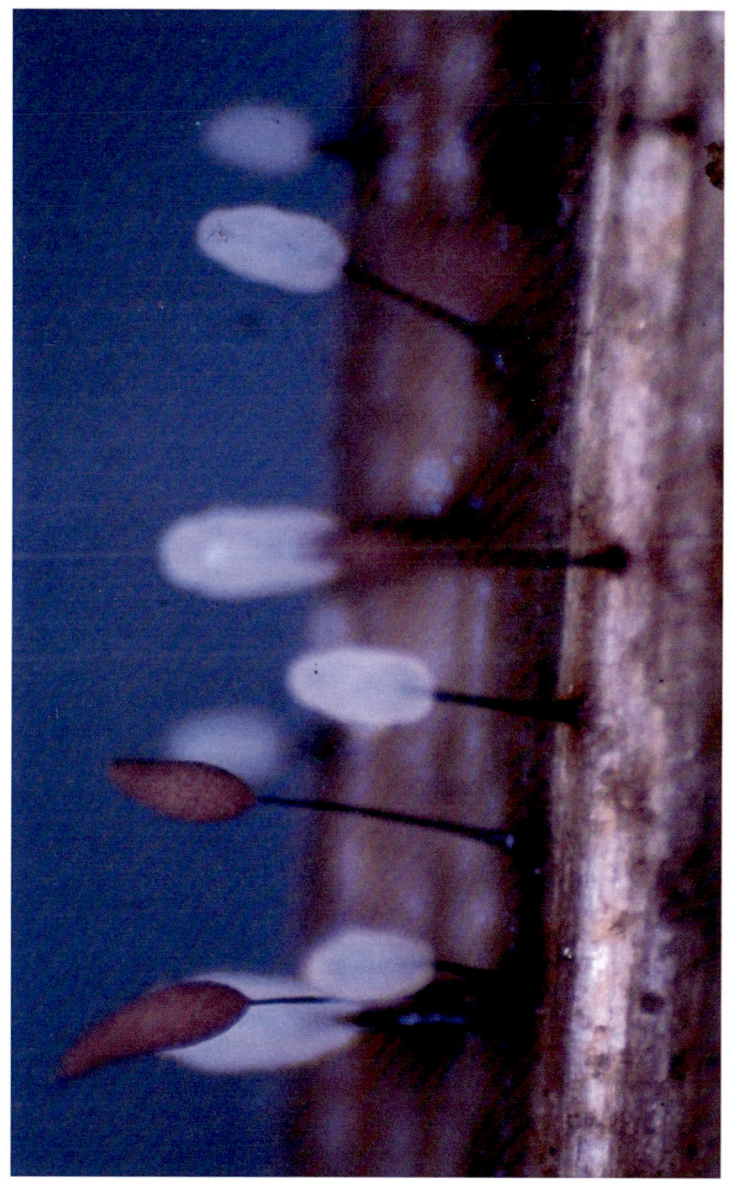

Figure 4: *Comatricha tenerrima* – a frequent species on dead herbaceous stems in damp places, and a favourite species of the distinguished mycologist, Douglas Graddon, who lived at Macclesfield (picture by P. Brustkern).

Figure 5: *Arcyria denudate* – a common species of stumps and fallen trunks, especially of beech and oak.

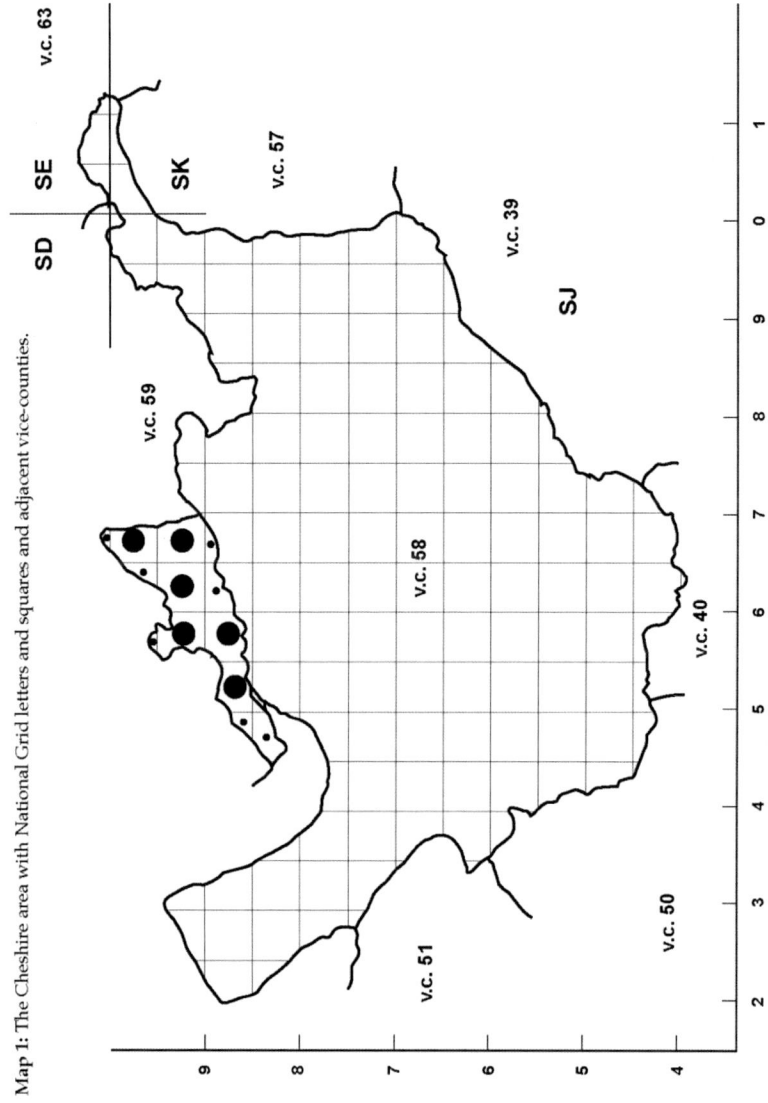

Map 1: The Cheshire area with National Grid letters and squares and adjacent vice-counties.

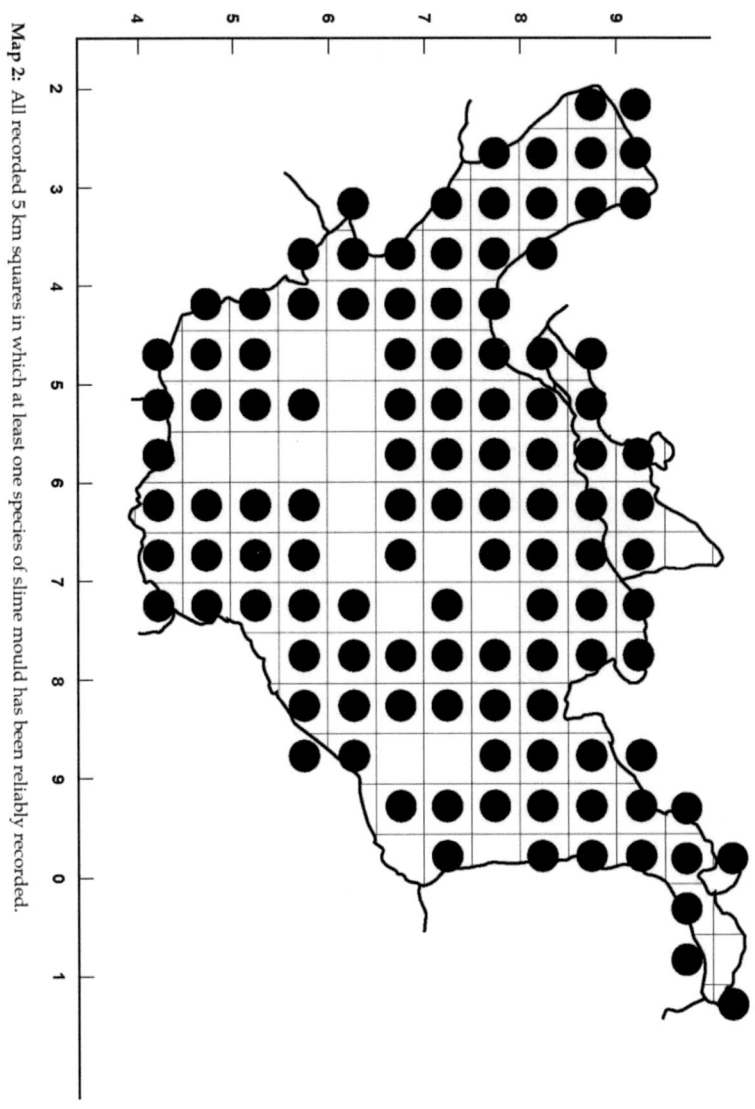

Map 2: All recorded 5 km squares in which at least one species of slime mould has been reliably recorded.

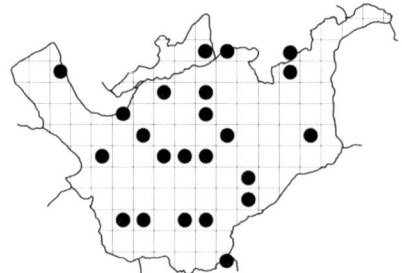

Map 3: The distribution of *Pocheina rosea*.

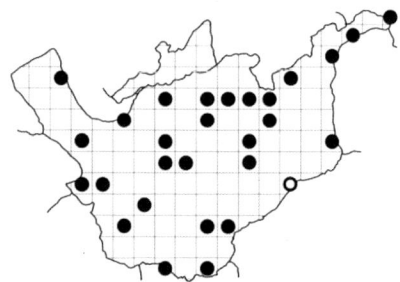

Map 4: *Ceratiomyxa fruticulosa.*

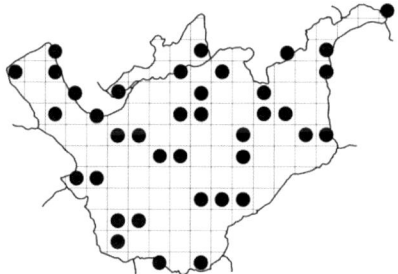

Map 5: *Echinostelium brooksii.*

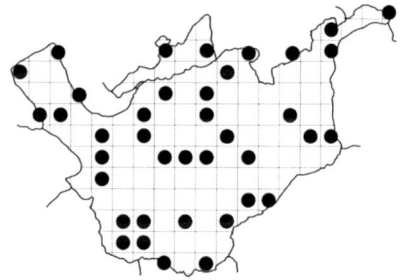

Map 6: *Echinostelium colliculosum.*

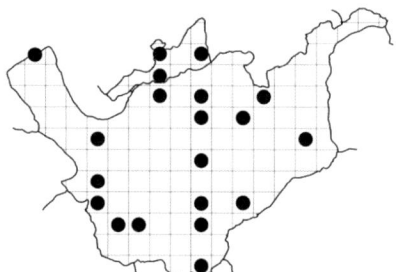

Map 7: *Echinostelium corynophorum.*

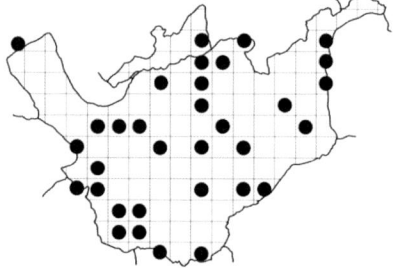

Map 8: *Echinostelium fragile.*

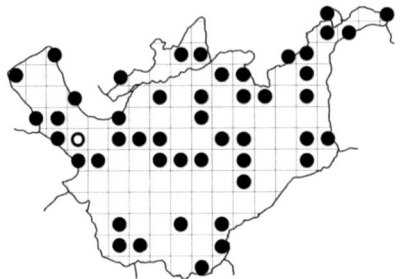

Map 9: *Echinostelium minutum.*

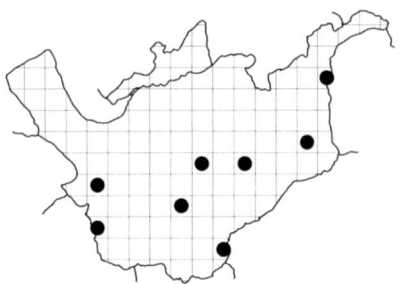

Map 10: *Licea belmontiana.*

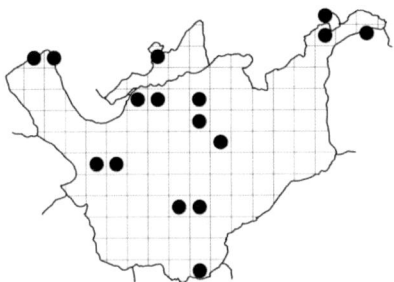

Map 11: *Licea biforis.*

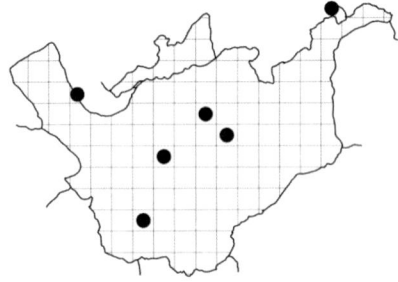

Map 12: *Licea bryophila.*

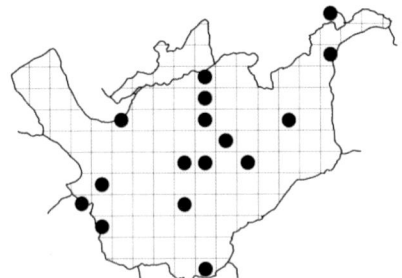

Map 13: *Licea clarkii*

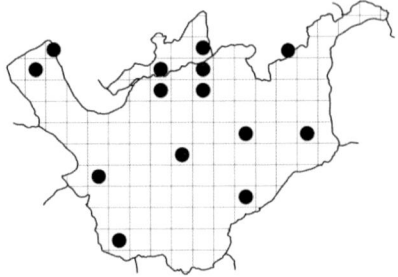

Map 14: *Licea denudescens.*

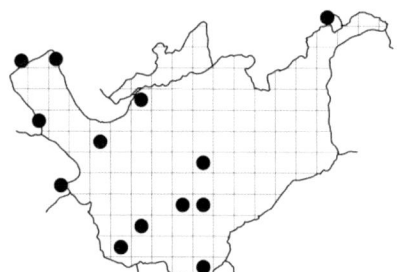

Map 15: *Licea eleanorae.*

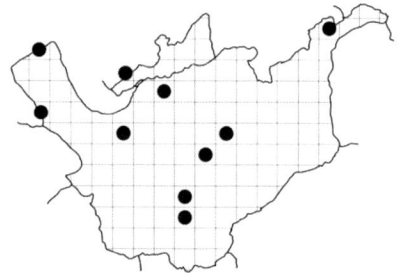

Map 16: *Licea erddigensis.*

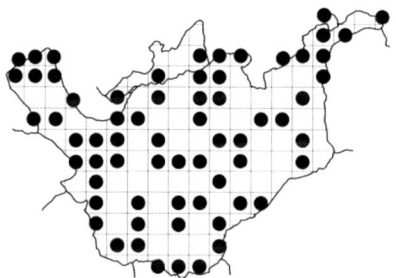

Map 17: *Licea kleistobolus.*

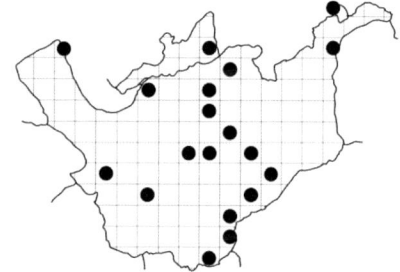

Map 18: *Licea marginata.*

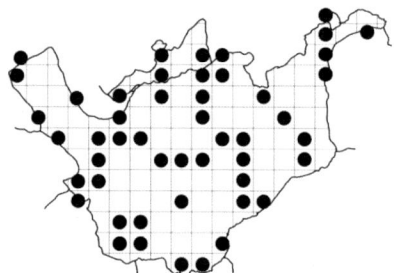

Map 19: *Licea microscopica.*

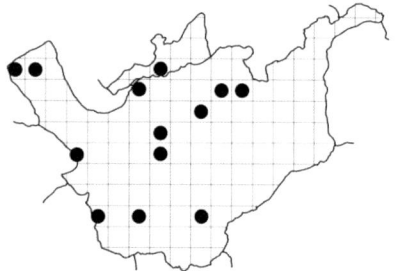

Map 20: *Licea minima.*

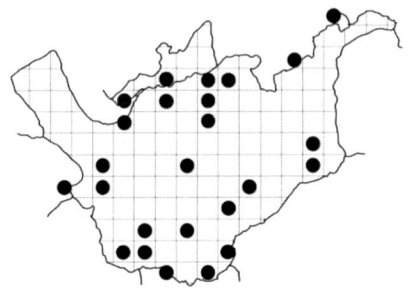

Map 21: *Licea operculata.*

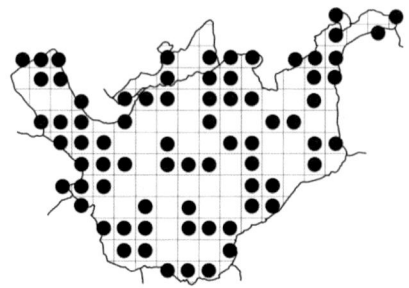

Map 22: *Licea parasitica.*

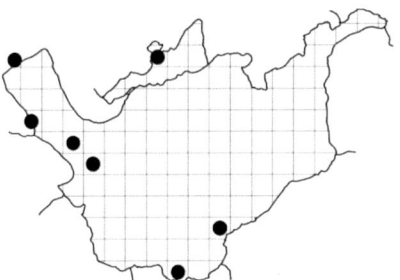

Map 23: *Licea pedicellata.*

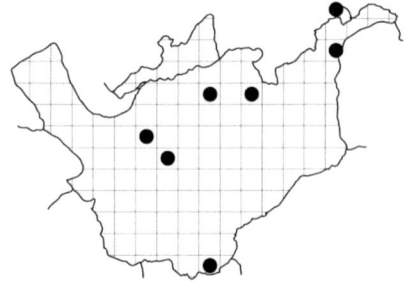

Map 24: *Licea pusilla.*

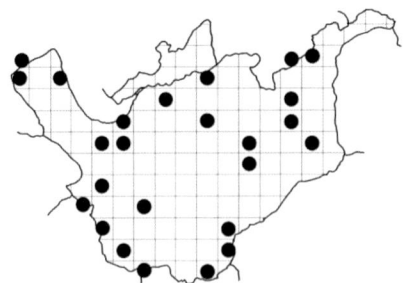

Map 25: *Licea pygmaea.*

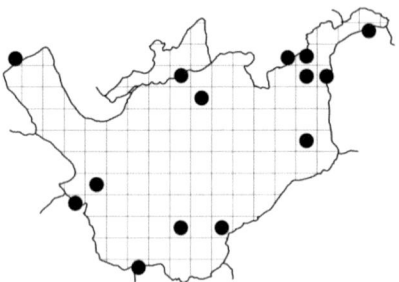

Map 26: *Licea scyphoides.*

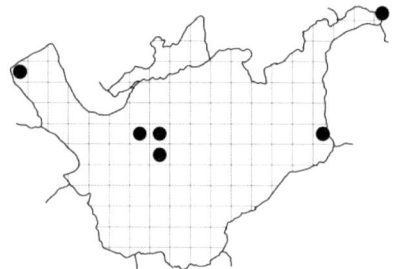

Map 27: *Licea variabilis.*

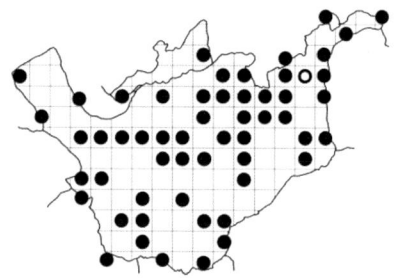

Map 28: *Lycogala epidendrum sensu lato.*

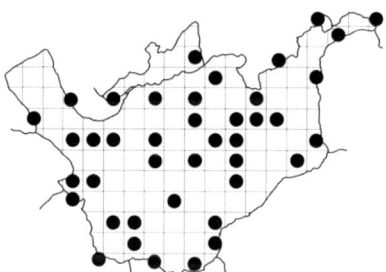

Map 29: *Lycogala terrestre.*

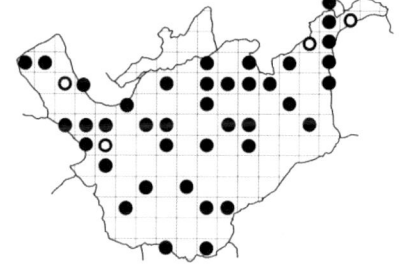

Map 30: *Reticularia lycoperdon.*

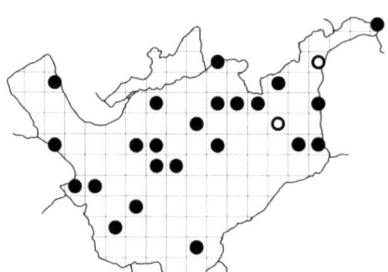

Map 31: *Tubifera ferruginosa.*

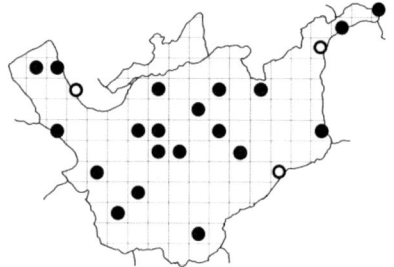

Map 32: *Cribraria argillacea.*

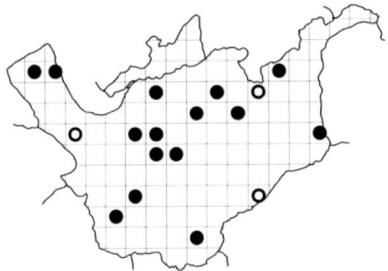

Map 33: *Cribraria aurantiaca.*

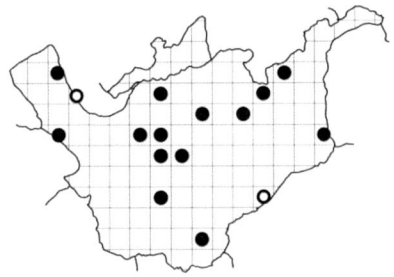

Map 34: *Cribraria cancellata.*

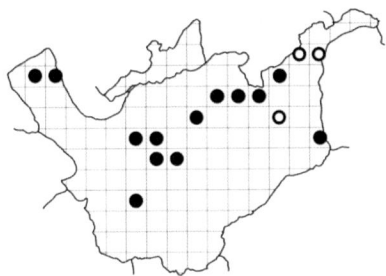

Map 35: *Cribraria rufa.*

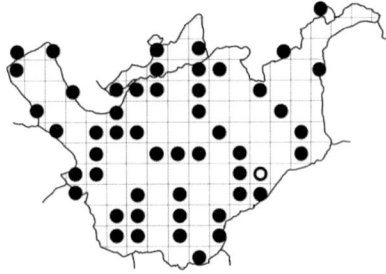

Map 36: *Calomyxa metallica.*

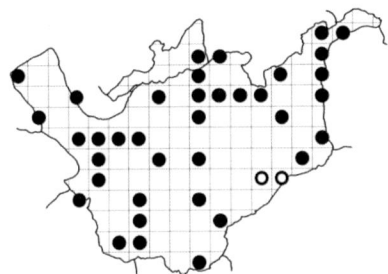

Map 37: *Arcyria cinerea.*

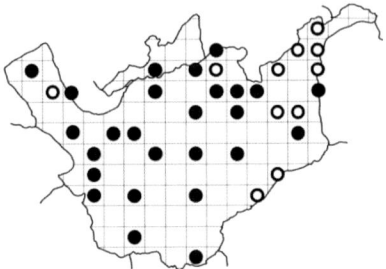

Map 38: *Arcyria denudata.*

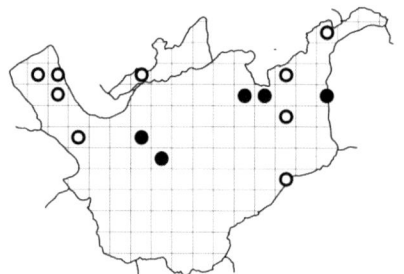

Map 39: *Arcyria ferruginea.*

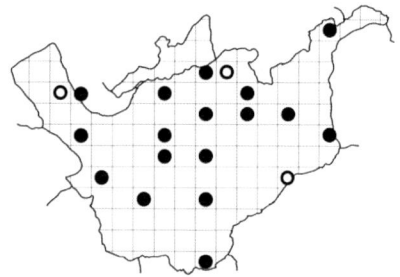

Map 40: *Arcyria incarnata.*

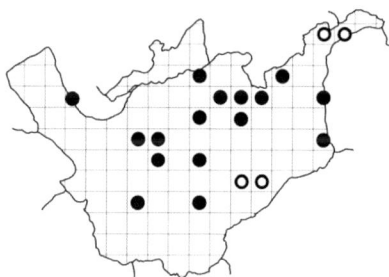

Map 41: *Arcyria obvelata.*

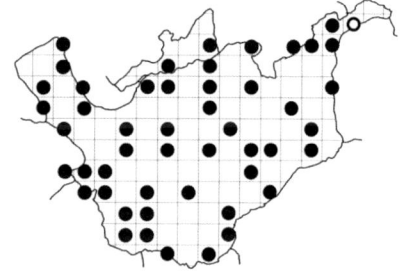

Map 42: *Arcyria pomiformis.*

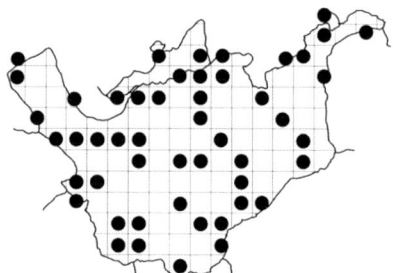

Map 43: *Perichaena chrysosperma.*

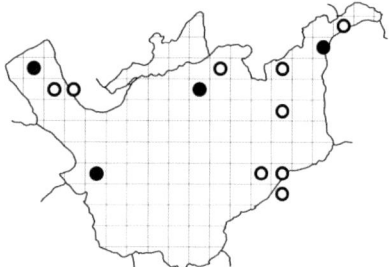

Map 44: *Perichaena corticalis.*

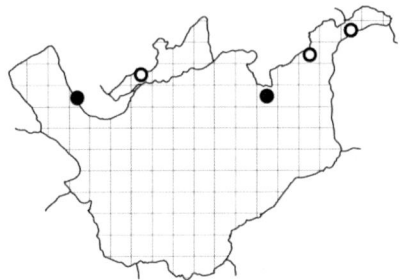

Map 45: *Perichaena depressa.*

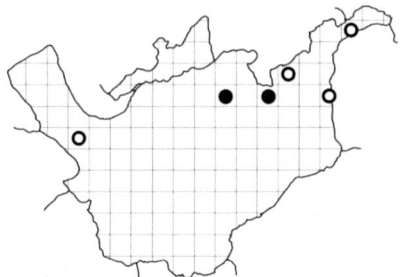

Map 46: *Hemitrichia calyculata.*

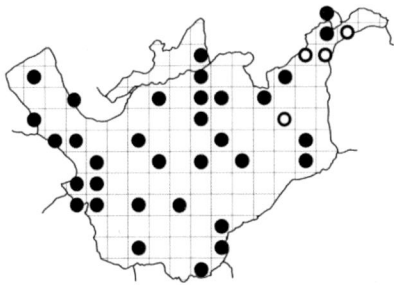

Map 47: *Metatrichia floriformis.*

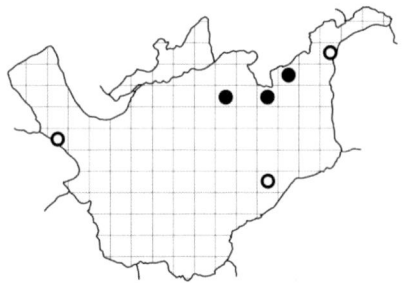

Map 48: *Metatrichia vesparium.*

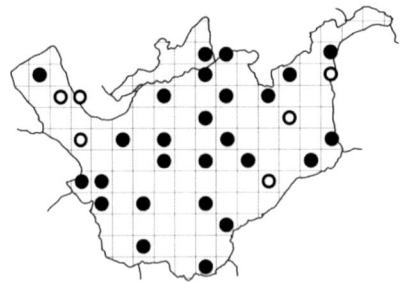

Map 49: *Trichia affinis.*

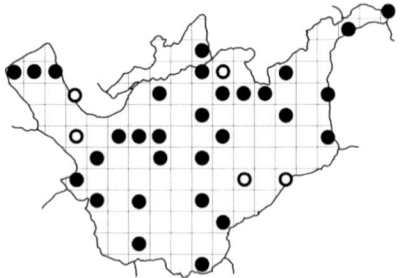

Map 50: *Trichia botrytis.*

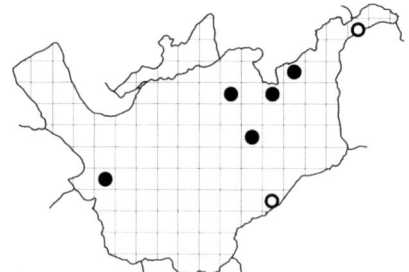

Map 51: *Trichia contorta.*

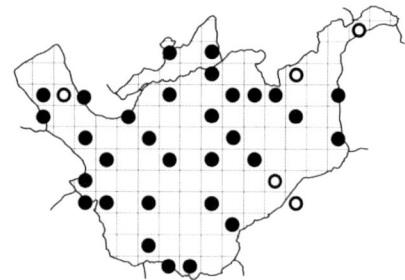

Map 52: *Trichia decipiens.*

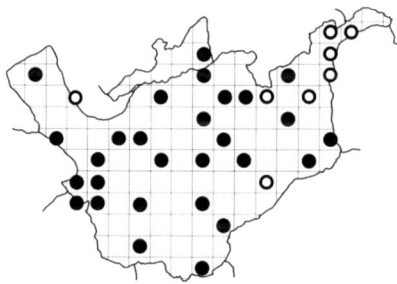

Map 53: *Trichia persimilis.*

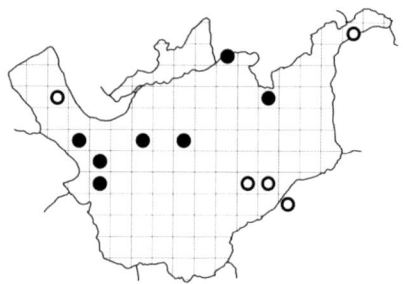

Map 54: *Trichia scabra.*

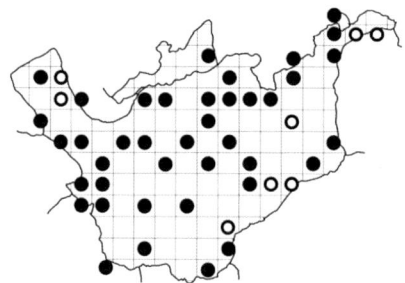

Map 55: *Trichia varia.*

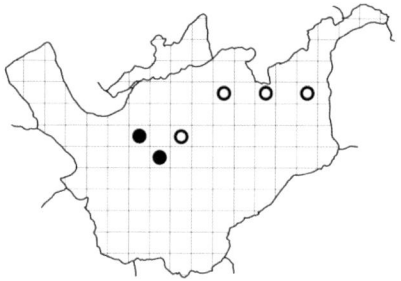

Map 56: *Amaurochaete atra.*

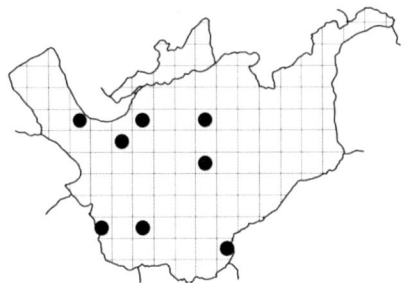

Map 57: *Comatricha laxa.*

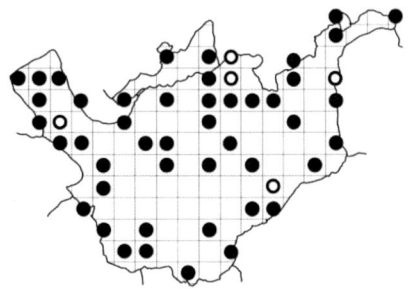

Map 58: *Comatricha nigra.*

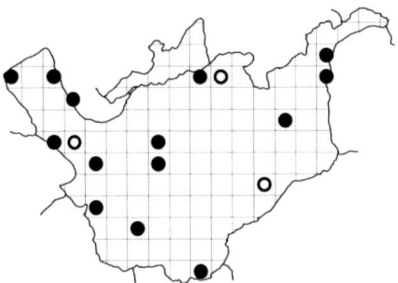

Map 59: *Comatricha pulchella.*

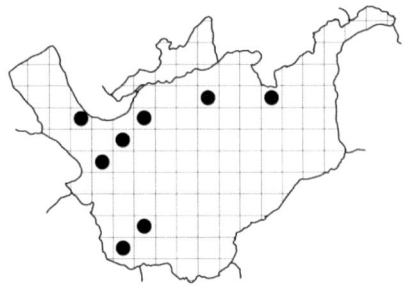

Map 60: *Comatricha rigidireta.*

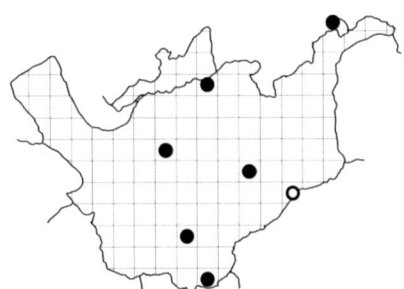

Map 61: *Comatricha tenerrima.*

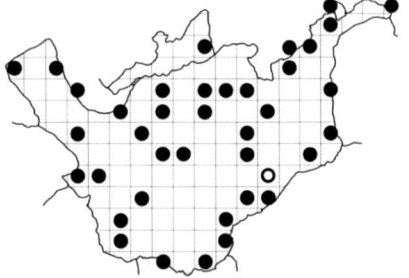

Map 62: *Enerthenema papillatum.*

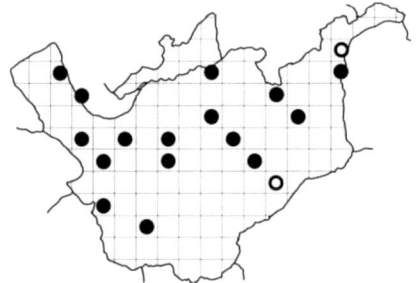

Map 63: *Lamproderma scintillans.*

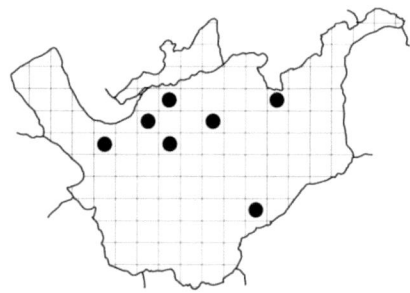

Map 64: *Paradiacheopsis cribrata.*

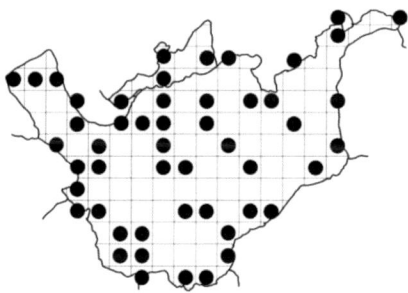

Map 65: *Paradiacheopsis fimbriata.*

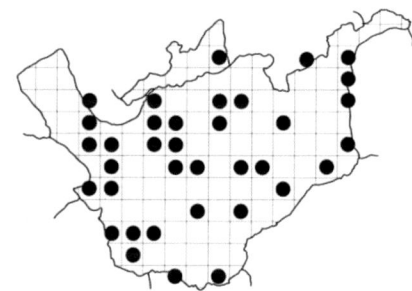

Map 66: *Paradiacheopsis solitaria.*

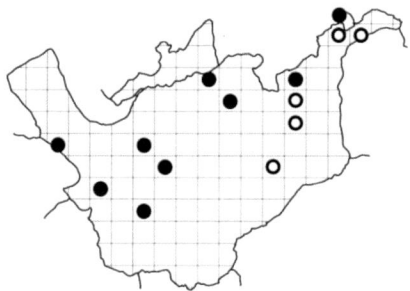

Map 67: *Stemonitis axifera.*

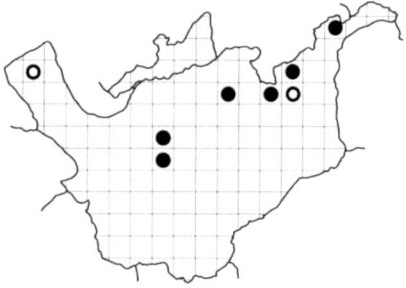

Map 68: *Stemonitis flavogenita.*

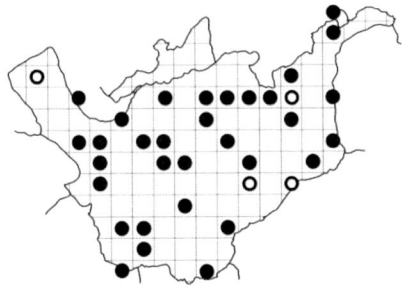

Map 69: *Stemonitis fusca.*

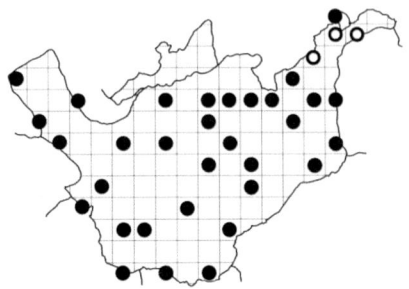

Map 70: *Stemonitopsis typhina.*

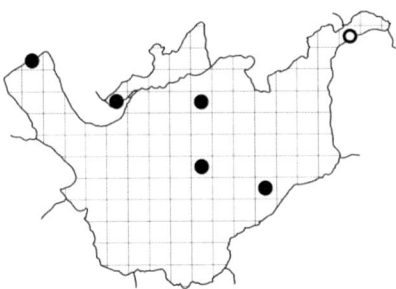

Map 71: *Badhamia affinis.*

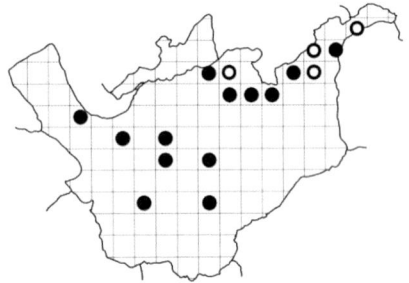

Map 72: *Badhamia panicea.*

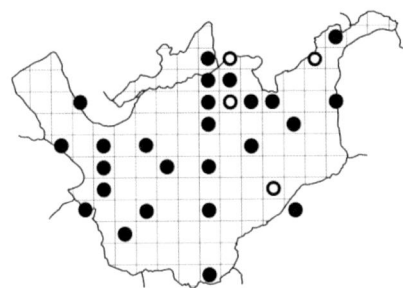

Map 73: *Badhamia utricularis.*

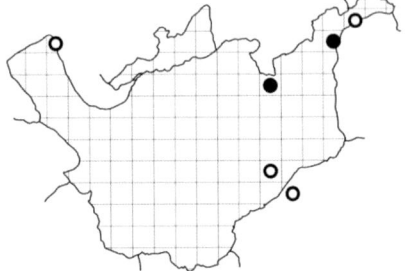

Map 74: *Craterium leucocephalum.*

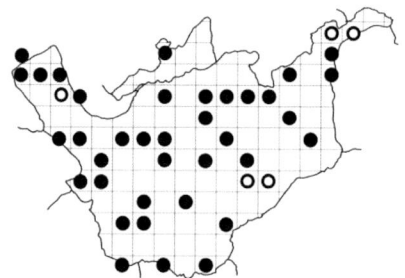

Map 75: *Craterium minutum.*

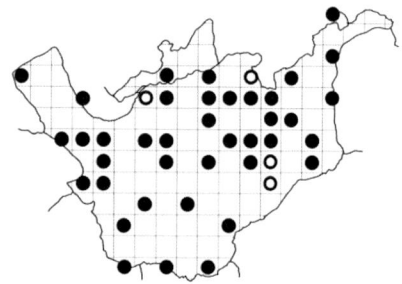

Map 76: *Fuligo septica.*

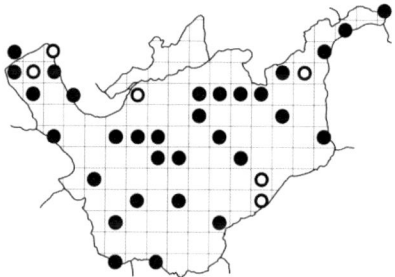

Map 77: *Leocarpus fragilis.*

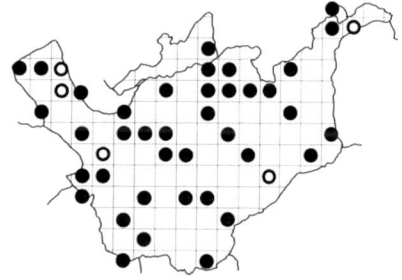

Map 78: *Physarum album.*

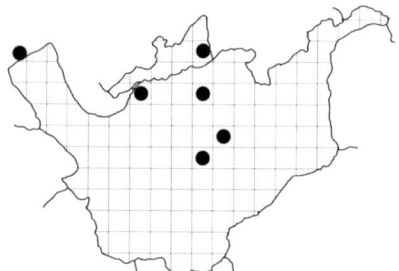

Map 79: *Physarum auriscalpium.*

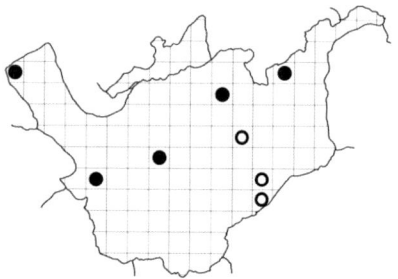

Map 80: *Physarum bitectum.*

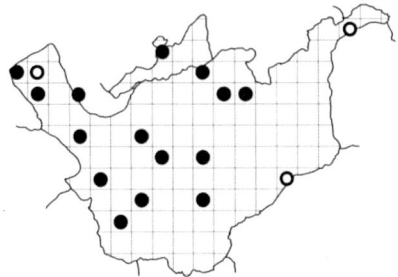

Map 81: *Physarum bivalve.*

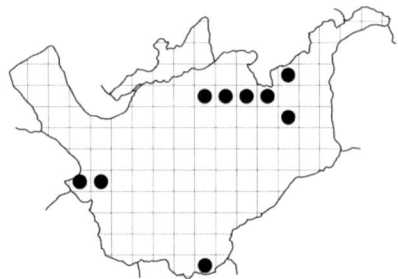

Map 82: *Physarum cinereum.*

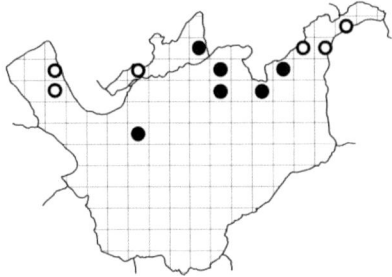

Map 83: *Physarum compressum.*

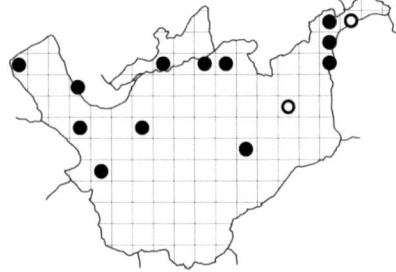

Map 84: *Physarum leucophaeum.*

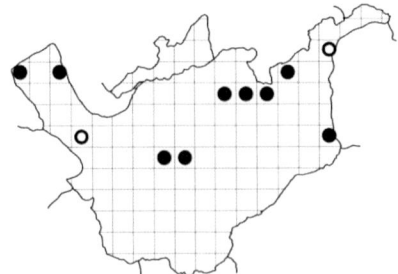

Map 85: *Physarum viride.*

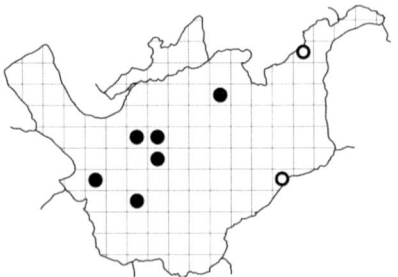

Map 86: *Diachea leucopodia.*

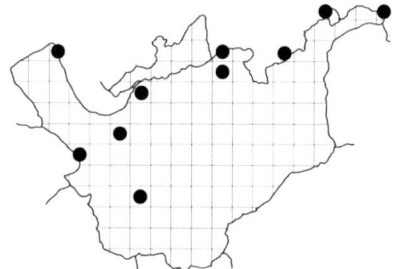

Map 87: *Diderma chondrioderma.*

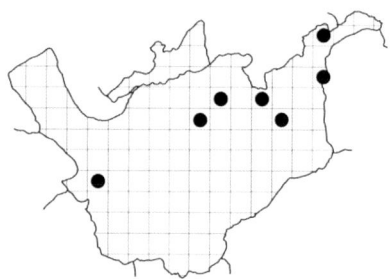

Map 88: *Diderma effusum.*

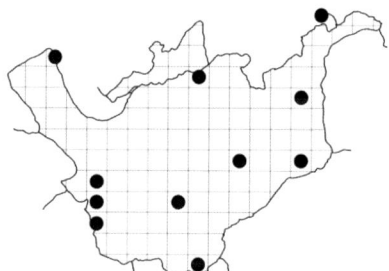

Map 89: *Diderma hemisphaericum.*

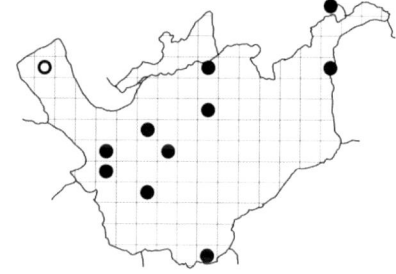

Map 90: *Didymium bahiense.*

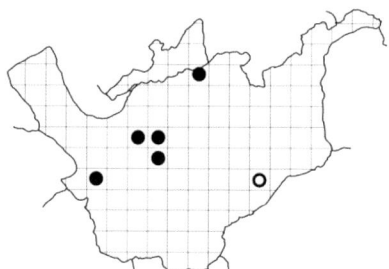

Map 91: *Didymium clavus.*

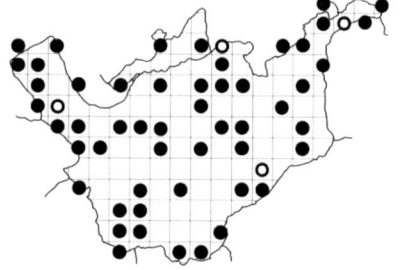

Map 92: *Didymium difforme.*

Biodiversity in the North West

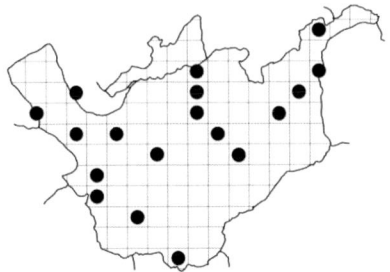

Map 93: *Didymium ilicinum.*

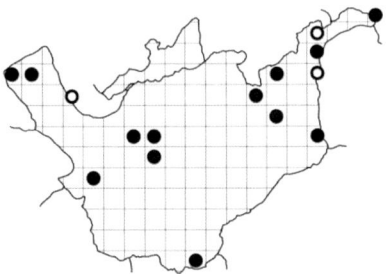

Map 94: *Didymium melanospermum.*

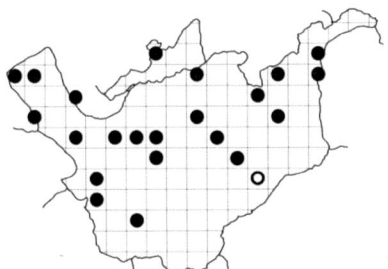

Map 95: *Didymium nigripes.*

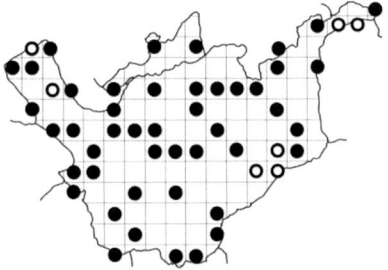

Map 96: *Didymium squamulosum.*

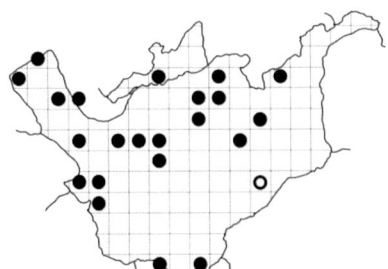

Map 97: *Mucilago crustacea.*